VOLUME	EDITOR-IN-CHIEF	PAGES	
34	William S. Johnson	121	*Out of print*
35	T. L. Cairns	122	*Out of print*
36	N. J. Leonard	120	*Out of print*
37	James Cason	109	*Out of print*
38	John C. Sheehan	120	*Out of print*
39	Max Tishler	114	*Out of print*

Collective Vol. IV   A revised edition of Annual Volumes 30–39
    Norman Rabjohn, *Editor-in-Chief*   1036

40	Melvin S. Newman	114	*Out of print*
41	John D. Roberts	118	*Out of print*
42	Virgil Boekelheide	118	*Out of print*
43	B. C. McKusick	124	*Out of print*
44	The Late William E. Parham	131	*Out of print*
45	William G. Dauben	118	*Out of print*
46	E. J. Corey	146	*Out of print*
47	William D. Emmons	140	*Out of print*
48	Peter Yates	164	*Out of print*
49	Kenneth B. Wiberg	124	*Out of print*

Collective Vol. V   A revised edition of Annual Volumes 40–49
    Henry E. Baumgarten, *Editor-in-Chief*   1234

Cumulative Indices to Collective Volumes I, II, III, IV, V
    Ralph L. and The Late Rachel
    H. Shriner, *Editors*

50	Ronald Breslow	136	*Out of print*
51	Richard E. Benson	209	*Out of print*
52	Herbert O. House	192	*Out of print*
53	Arnold Brossi	193	*Out of print*
54	Robert E. Ireland	155	*Out of print*
55	Satoru Masamune	150	*Out of print*
56	George H. Buchi	144	*Out of print*
57	Carl R. Johnson	135	*Out of print*
58	The Late William A. Sheppard	216	*Out of print*
59	Robert M. Coates	267	*Out of print*

Collective Vol. VI   A revised edition of Annual Volumes 50–59
    Wayland E. Noland, *Editor-in-Chief*
    In preparation

60	Orville L. Chapman		
61	The Late Robert V. Stevens		
62	Martin F. Semmelhack	269	
63	Gabriel Saucy	291	
64	Andrew S. Kende	308	
65	Edwin Vedejs	278	

ORGANIC SYNTHESES

ORGANIC SYNTHESES

AN ANNUAL PUBLICATION OF SATISFACTORY
METHODS FOR THE PREPARATION
OF ORGANIC CHEMICALS

VOLUME 65
1987

BOARD OF EDITORS

EDWIN VEDEJS, *Editor-in-Chief*

CLAYTON H. HEATHCOCK	K. BARRY SHARPLESS
ALBERT I. MEYERS	BRUCE E. SMART
RYOJI NOYORI	JAMES D. WHITE
LARRY E. OVERMAN	EKKEHARD WINTERFELDT
LEO A. PAQUETTE	

THEODORA W. GREENE, *Assistant Editor*
JEREMIAH P. FREEMAN, *Secretary to the Board*
Department of Chemistry, University of Notre Dame,
Notre Dame, Indiana 46556

ADVISORY BOARD

RICHARD T. ARNOLD	ALBERT ESCHENMOSER	WAYLAND E. NOLAND
HENRY E. BAUMGARTEN	IAN FLEMING	CHARLES C. PRICE
RICHARD E. BENSON	E. C. HORNING	NORMAN RABJOHN
VIRGIL BOEKELHEIDE	HERBERT O. HOUSE	JOHN D. ROBERTS
RONALD BRESLOW	ROBERT E. IRELAND	GABRIEL SAUCY
ARNOLD BROSSI	CARL R. JOHNSON	R. S. SCHREIBER
GEORGE H. BÜCHI	WILLIAM S. JOHNSON	DIETER SEEBACH
T. L. CAIRNS	ANDREW S. KENDE	MARTIN F. SEMMELHACK
JAMES CASON	N. J. LEONARD	JOHN C. SHEEHAN
ORVILLE L. CHAPMAN	B. C. MCKUSICK	RALPH L. SHRINER
ROBERT M. COATES	C. S. MARVEL	H. R. SNYDER
E. J. COREY	SATORU MASAMUNE	MAX TISHLER
WILLIAM G. DAUBEN	WATARU NAGATA	KENNETH B. WIBERG
WILLIAM D. EMMONS	MELVIN S. NEWMAN	PETER YATES

FORMER MEMBERS OF THE BOARD, NOW DECEASED

ROGER ADAMS	ARTHUR C. COPE	OLIVER KAMM
HOMER ADKINS	NATHAN L. DRAKE	C. R. NOLLER
C. F. H. ALLEN	L. F. FIESER	W. E. PARHAM
WERNER E. BACHMANN	R. C. FUSON	WILLIAM A. SHEPPARD
A. H. BLATT	HENRY GILMAN	LEE IRVIN SMITH
WALLACE H. CAROTHERS	CLIFF S. HAMILTON	ROBERT V. STEVENS
H. T. CLARKE	W. W. HARTMAN	FRANK C. WHITMORE
J. B. CONANT	JOHN R. JOHNSON	

JOHN WILEY & SONS

NEW YORK · CHICHESTER · BRISBANE · TORONTO · SINGAPORE

Published by John Wiley & Sons, Inc.

Copyright © 1987 by Organic Syntheses, Inc.

All rights reserved. Published simultaneously in Canada.

Reproduction or translation of any part of this work beyond that permitted by Section 107 or 108 of the 1976 United States Copyright Act without the permission of the copyright owner is unlawful.

"John Wiley & Sons, Inc. is pleased to publish this volume of Organic Syntheses on behalf of Organic Syntheses, Inc. Although Organic Syntheses, Inc. has assured us that each preparation contained in this volume has been checked in an independent laboratory and that any hazards that were uncovered are clearly set forth in the write-up of each preparation, John Wiley & Sons, Inc. does not warrant the preparations against any safety hazards and assumes no liability with respect to the use of the preparations."

Library of Congress Catalog Card Number: 21-17747
ISBN 0-471-63637-1

Printed in the United States of America

10 9 8 7 6 5 4 3 2 1

NOTICE

With Volume 62, the Editors of *Organic Syntheses* began a new presentation and distribution policy to shorten the time between submission and appearance of an accepted procedure. The soft cover edition of this volume is produced by a rapid and inexpensive process, and is sent at no charge to members of the Organic Division of the American Chemical Society. The soft cover edition is intended as the personal copy of the owner and is not for library use. A hard cover edition is published by John Wiley and Sons, Inc. in the traditional format, and differs in content primarily in the inclusion of an index. The hard cover edition is intended primarily for library collections and is available for purchase through the publisher. Annual Volumes 60–64 will be included in a new five-year version of the collective volumes of *Organic Syntheses* which will appear as *Collective Volume Seven* in the traditional hard cover format, after the appearance of annual volume 64. It will be available for purchase from the publishers. The Editors hope that the new *Collective Volume* series, appearing twice as frequently as the previous decennial volumes, will provide a permanent and timely edition of the procedures for personal and institutional libraries. The Editors welcome comments and suggestions from users concerning the new editions.

NOMENCLATURE

Both common and systematic names of compounds are used throughout this volume, depending on which the Editor-in-Chief felt was more appropriate. The *Chemical Abstracts* indexing name for each title compound, if it differs from the title name, is given as a subtitle. Systematic *Chemical Abstracts* nomenclature, used in both the 9th and 10th Collective Indexes for the title compound and a selection of other compounds mentioned in the procedure, is provided in an appendix at the end of each preparation. Registry numbers, which are useful in computer searching and identification, are also provided in these appendixes. Whenever two names are concurrently in use and one name is the correct *Chemical Abstracts* name, that name is adopted. For example, both diethyl ether and ethyl ether are normally used. Since ethyl ether is the established *Chemical Abstracts* name for the 8th Collective Index, it has been used in this volume. The 9th Collective Index name is 1,1'-oxybisethane, which the Editors consider too cumbersome.

SUBMISSION OF PREPARATIONS

Organic Syntheses welcomes and encourages submission of experimental procedures which lead to compounds of wide interest or which illustrate important new developments in methodology. The Editorial Board will consider proposals in outline format as shown below, and will request full experimental details for those proposals which are of sufficient interest. Submissions which are longer than three steps from commercial sources or from existing *Organic Syntheses* procedures will be accepted only in unusual circumstances.

Organic Syntheses Proposal Format
1. Authors
2. Literature reference or enclose preprint if available.
3. Proposed sequence
4. Best current alternative(s)

5. a. Proposed scale, final product:
   b. Overall yield:
   c. Method of isolation and purification:
   d. Purity of product (%):
   e. How determined?
6. Any unusual apparatus or experimental technique:
7. Any hazards?
8. Source of starting material?
9. Utility of method or usefulness of product.

Submit to: Dr. Jeremiah P. Freeman, Secretary
           Department of Chemistry
           University of Notre Dame
           Notre Dame, IN 46556

Proposals will be evaluated in outline form, again after submission of full experimental details and discussion, and finally, by checking experimental procedures. A form that details the preparation of a complete procedure (Notice to Submitters) may be obtained from the Secretary.

Additions, corrections, and improvements to the preparations previously published are welcomed; these should be directed to the Secretary. However, checking of such improvements will only be undertaken when new methodology is involved. Substantially improved procedures have been included in the Collective Volumes in place of a previously published procedure.

PREFACE

This volume reflects intensive activity in several areas of synthetic organic chemistry. The volume begins with enol/carbonyl condensations and several variants for synthesis of 5-membered carbocycles. Next comes a series of examples illustrating the synthesis of alkynes, alkenes, and aromatic carbocycles. Heteroelement chemistry is featured in the extensive use of organosilicon reagents, in several applications of transition element chemistry, in reactive nitrogen intermediates, and in the preparation of β-lactams, indoles, and other heterocycles. The last and largest section of the volume deals with a variety of chiral auxiliaries which are important in asymmetric synthesis. This field is in the midst of explosive growth and will undoubtedly see major changes and improvements in techniques and results over the next decade.

I would like to thank Carole Klingbeil, Joyce Bohling, and Professor Jeremiah P. Freeman and his office staff for their extensive help and patience in the preparation of text for this volume. Mr. P. Kierkus deserves credit for the structures, drawn with the ChemDraw™ program and for transforming a great deal of rough copy into final diagrams. Lastly, my thanks to those of my students over the past several years who have spent their time and energy checking procedures and contributing in other ways which left me time for this volume.

E. VEDEJS

Madison, Wisconsin
January 1987

A. HAROLD BLATT
January 9, 1903–March 19, 1986

A. Harold Blatt, second Secretary to the Board of Editors of *Organic Syntheses, Inc.*, 1938-1943, and co-editor with Henry Gilman of the revised edition of Collective Volume 1 of *Organic Syntheses,* passed away on March 19, 1986 in Melbourne, Florida at the age of 83.

Dr. Blatt was born in Cincinnati, Ohio and received B.S., M.A., and Ph.D. degrees from Harvard University in 1923-1926. He held postdoctoral positions at the College de France in Paris, Harvard University, and the University of Buffalo before he joined the faculty at Howard University as an associate professor in 1932. He became a member of the newly-formed Queens College in 1939, where he was a professor, and stayed for 32 years. His academic pursuits were interrupted during World War II when he was a Science Liaison Officer, the London Mission, in the Office of Scientific Research and Development (1944-1945), and a Technical Aide to Division 8 during the latter year.

Dr. Blatt also edited Collective Volume 2 of *Organic Syntheses* (1943), and served for many years on the Board of Directors, where his expertise and knowledge of finance were of inestimable value. At corporation meetings, his versatility was shown by the skill he demonstrated in the selection of the dinner wines. His editorial expertise was used also by Organic Reactions, Inc. where he was a member of the Editorial Board (1948-1954), and then served on the Advisory Board until 1986.

Dr. Blatt's teaching and research activities covered the period from his Harvard days into the 1980's. He was a co-editor with James B. Conant of the well-known text, *The Chemistry of Organic Compounds,* 3rd edition, which was published in 1947, and used widely during the 1950's. The text offered a new approach for organic chemistry students to the subjects of reaction rates and equilibria. It also presented new physiochemical concepts and data, as well as an effort to cover some of the major topics of biochemistry and pharmacology, and relate their chemistry to the principles expounded in the book.

Harold Blatt served for three terms as chairman of the Chemistry Department at Queens College beginning in 1961, retired in 1971, and then moved to Melbourne, Florida where he was a member of the Chemistry Department at the Florida Institute of Technology.

Dr. Joel H. Blatt, his son, and a sister survive Dr. Blatt.

NORMAN RABJOHN

September 19, 1986

CONTENTS

Nigel D. A. Walshe, Graham B. T. Goodwin, Graham C. Smith, and Frank E. Woodward	1	ACETONE TRIMETHYLSILYL ENOL ETHER
Teruaki Mukaiyama and Koichi Narasaka	6	3-HYDROXY-3-METHYL-1-PHENYL-1-BUTANONE BY CROSSED ALDOL REACTION
Koichi Narasaka	12	3-DIMETHYL-1,5-DIPHENYLPENTANE-1,5-DIONE
Eiichi Nakamura and Isao Kuwajima	17	RING EXPANSION AND CLEAVAGE OF SUCCINOIN DERIVATIVES: SPIRO[4.5]DECANE-1,4-DIONE AND ETHYL 4-CYCLOHEXYL-4-OXOBUTANOATE
H. Stetter, H. Kuhlmann and W. Haese	26	THE STETTER REACTION: 3-METHYL-2-PENTYL-2-CYCLOPENTEN-1-ONE (DIHYDROJASMONE)
Dale L. Boger, Christine E. Brotherton, and Gunda I. Georg	32	PREPARATION AND THREE-CARBON + TWO-CARBON CYCLOADDITION OF A CYCLOPROPENONE KETAL: CYCLOPROPENONE 1,3-PROPANEDIOL KETAL
Richard S. Threlkel, John E. Bercaw, Paul F. Seidler, Jeffrey M. Stryker, and Robert G. Bergman	42	1,2,3,4,5-PENTAMETHYLCYCLOPENTADIENE
M. Olomucki and J. Y. Le Gall	47	ALKOXYCARBONYLATION OF PROPARGYL CHLORIDE: METHYL 4-CHLORO-2-BUTYNOATE
Graham E. Jones, David A. Kendrick, and Andrew B. Holmes	52	1,4-BIS(TRIMETHYLSILYL)BUTA-1,3-DIYNE
Andrew B. Holmes and Chris N. Sporikou	61	TRIMETHYLSILYLACETYLENE
Anna Bou, Miquel A. Pericàs, Antoni Riera and Fèlix Serratosa	68	DIALKOXYACETYLENES: DI-tert-BUTOXYETHYNE, A VALUABLE SYNTHETIC INTERMEDIATE
Luciano Lombardo	81	METHYLENATION OF CARBONYL COMPOUNDS: (+)-3-METHYLENE-cis-p-MENTHANE

Eric Block and Mohammad Aslamb	90	A GENERAL SYNTHETIC METHOD FOR THE PREPARATION OF CONJUGATED DIENES FROM OLEFINS USING BROMOMETHANESULFONYL BROMIDE: 1,2-DIMETHYLENECYCLOHEXANE
Dale L. Boger and Michael D. Mullican	98	PREPARATION AND INVERSE ELECTRON DEMAND DIELS-ALDER REACTION OF AN ELECTRON-DEFICIENT DIENE: METHYL 2-OXO-5,6,7,8-TETRAHYDRO-2H-1-BENZOPYRAN-3-CARBOXYLATE
F. E. Ziegler, K. W. Fowler, W. B. Rodgers, and R. T. Wester	108	AMBIENT TEMPERATURE ULLMANN REACTION: 4,5,4',5'-TETRAMETHOXY-1,1'-BIPHENYL-2,2'-DICARBOXALDEHYDE
Steven K. Davidsen, Gerald W. Phillips, and Stephen F. Martin	119	GEMINAL ACYLATION-ALKYLATION AT A CARBONYL CENTER USING DIETHYL N-BENZYLIDENEAMINOMETHYLPHOSPHONATE: PREPARATION OF 2-METHYL-2-PHENYL-4-PENTENAL
Stuart J. Mickel, and modified by Chi-Nung Hsiao and Marvin J. Miller	135	4-ACETOXYAZETIDIN-2-ONE: SYNTHESIS OF A KEY BETA-LACTAM INTERMEDIATE BY A [2 + 2] CYCLOADDITION ROUTE
Louis S. Hegedus, Michael A. McGuire, and Lisa M. Schultze	140	1,3-DIMETHYL-3-METHOXY-4-PHENYL-AZETIDINONE
Jan Bergman and Peter Sand	146	4-NITROINDOLE
J. Buter and Richard M. Kellogg	150	SYNTHESIS OF MACROCYCLIC SULFIDES USING CESIUM THIOLATES: 1,4,8,11-TETRATHIACYCLOTETRADECANE
Günter Kresze, Hans Braxmeier, and Heribert Münsterer	159	ALLYLCARBAMATES BY THE AZA-ENE REACTION: METHYL N-(2-METHYL-2-BUTENYL)CARBAMATE
E. J. Corey and Andrew W. Gross	166	tert-BUTYL-tert-OCTYLAMINE
Dieter Enders, Peter Fey, and Helmut Kipphardt	173	(S)-(-)-1-AMINO-2-METHOXYMETHYL-PYRROLIDINE (SAMP) AND (R)-(+)-1-AMINO-2-METHOXYMETHYLPYRROLIDINE (RAMP), VERSATILE CHIRAL AUXILIARIES
Dieter Enders, Helmut Kipphardt, and Peter Fey	183	ASYMMETRIC SYNTHESES USING THE SAMP-/RAMP-HYDRAZONE METHOD: (S)-(+)-4-METHYL-3-HEPTANONE
Oswald Ort	203	(-)-8-PHENYLMENTHOL

Ernest L. Eliel, Joseph E. Lynch, Fumitaka Kume, and Stephen V. Frye	215	CHIRAL 1,3-OXATHIANE FROM (+)-PULEGONE: HEXAHYDRO-4,4,7-TRIMETHYL-4H-1,3-BENZOXATHIIN
Charles A. Brown and Prabhakav K. Jadhav	224	(-)-α-PINENE BY ISOMERIZATION OF (-)-β-PINENE
René Imwinkelried, Martin Schiess, and Dieter Seebach	230	DIISOPROPYL (2S,3S)-2,3-O-ISOPROPYLIDENETARTRATE
Bernd Giese, J. Dupuis, and Marianne Nix	236	1-DEOXY-2,3,4,6-TETRA-O-ACETYL-1-(2-CYANOETHYL)-α-D-GLUCOPYRANOSE
S. Hanessian	243	6-BROMO-6-DEOXY HEXOSE DERIVATIVES BY RING-OPENING OF BENZYLIDENE ACETALS WITH N-BROMOSUCCINIMIDE: METHYL 4-O-BENZOYL-6-BROMO-6-DEOXY-α-D-GLUCOPYRANOSIDE
Unchecked Procedures	251	
Cumulative Author Index for Volume 65	256	
Cumulative Subject Index for Volume 65	258	

ORGANIC SYNTHESES

ACETONE TRIMETHYLSILYL ENOL ETHER

(Silane, trimethyl[(1-methylethenyl)oxy]-)

$$CH_3COCH_3 + (CH_3)_3SiCl + NaI \xrightarrow[CH_3CN]{(C_2H_5)_3N} CH_2=\underset{\underset{CH_3}{|}}{C}-OSi(CH_3)_3$$

Submitted by Nigel D. A. Walshe, Graham B. T. Goodwin, Graham C. Smith, and Frank E. Woodward.[1]
Checked by V. A. Palaniswamy and James D. White.

1. Procedure

To a 5-L, four-necked flask, equipped with a mechanical stirrer, reflux condenser with nitrogen inlet, thermometer and pressure-equalizing dropping funnel, are added 150 g (2.6 mol) of acetone (Note 1) and 192 g (1.9 mol) of triethylamine (Note 2) under a nitrogen atmosphere. To this mixture, stirred at room temperature under nitrogen, is added via the dropping funnel 200 g (1.84 mol) of chlorotrimethylsilane over 10 min (Note 3). The flask is then immersed in a waterbath and the contents are warmed to 35°C. The waterbath is removed and the dropping funnel is charged with a solution of 285 g (1.9 mol) of sodium iodide (Note 4) in 2.14 L of acetonitrile (Note 5). This solution is added to the stirred mixture in the flask at such a rate that the temperature of the reaction is maintained at 35-40°C without external heating or cooling (Note 6). The addition requires approximately 1 hr. When addition is complete, the reaction mixture is stirred for a further 2 hr at room temperature. The contents of the flask are then poured into 5 L of ice-cold water, and the aqueous mixture is extracted with two 1-L portions of pentane, and once with 500 mL of pentane. The combined pentane extracts are dried over anhydrous potassium carbonate, and filtered into a 3-L, round-bottomed flask. This is arranged for distillation at atmospheric pressure, incorporating a 30-cm Vigreux fractionating column. The pentane is distilled off at atmospheric pressure, until a head temperature of 88°C is attained. The crude material is transferred to a 500-mL flask, and the product is then distilled at atmospheric pressure through a 20-cm Vigreux column. A forerun of 20 g is collected between room temperature and 94°C. The product is the fraction boiling at 94-96°C, the yield of which is 116-130 g (48-54%) (Note 7).

2. Notes

1. "AnalaR" grade acetone, as supplied by BDH, was used.

2. Triethylamine was dried over potassium hydroxide pellets for at least 24 hr.

3. Commercial chlorotrimethylsilane was used without purification. When it was added to the acetone/triethylamine mixture, only a very mild exothermic reaction occurred (ca. 2°C). Dense white fumes formed, and a turbid solution was obtained.

4. Sodium iodide was reagent grade. It is essential to dry this material thoroughly. Heating at 140°C for 8 hr under reduced pressure (ca. 20 mm) is satisfactory. The loss of weight on drying is roughly 5%. If this is not done, hexamethyldisiloxane is the principal product.

5. Acetonitrile was reagent grade, dried by passage through 1 kg of neutral alumina (grade 1), and then stored over 3 Å molecular sieves.

6. A copious white precipitate forms at this stage. If the reaction is not mildly exothermic, then very poor yields of product are obtained.

7. The yield is based on chlorotrimethylsilane. Two small-scale runs - 0.124 mol and 0.37 mol, also based on chlorotrimethylsilane - gave yields of 60% and 61%, respectively, which the submitters also reported on the larger scale. The material from the large-scale run was 92% pure by gas-chromatographic analysis. The impurities, identified by NMR, are triethylamine (0.5%) and hexamethyldisiloxane (7.5%). The product has the following spectral characteristics; IR (film) cm^{-1}: 1650, 1280, 1260, 1050; ^{1}H NMR (CDCl$_3$) δ: 0.13 (s, 9 H, SiC$\underline{H}_3$), 1.69 (br s, 3 H, =CC$\underline{H}_3$), 3.92 (m, 2 H, =C$\underline{H}_2$).

3. Discussion

Trimethylsilyl enolates of aldehydes and ketones are now established as highly useful synthetic intermediates.[2] In particular, their Lewis acid-catalyzed reactions - e.g., alkylation[3] and mild, regiospecific aldol condensations[4] - provide useful alternatives to classical, base-generated metal enolate chemistry. This new methodology would be ideal for the introduction of the commonly-encountered acetonyl residue. However the required silyl enol ether of acetone is not commercially available, nor is a simple, reliable and economical synthesis adequately described in the literature. The above procedure is an adaptation of a literature method,[5] and relies on the generation of iodotrimethylsilane in situ. We have found that the precautions described in Notes 4 and 6 are crucial to the success of the preparation. This procedure makes available a useful reagent by a cheap, reliable route, starting from readily available materials, and in large or small quantity. The trimethylsilyl enol ether of acetone has been prepared previously in good yield by reaction of acetone with trimethylsilyl triflate and triethylamine.[6] However, the silyl triflate reagent is expensive for large-scale work. Another route[7] involves the mercuric iodide-catalyzed rearrangement of α-trimethylsilylacetone (obtained from trimethylsilylmethylmagnesium chloride and acetic anhydride). This is a laborious, low-yield process. Other methods include a synthesis from acetone, chlorotrimethylsilane, and triethylamine[8] (yields and exact procedure unspecified); or reaction of acetone with hexamethyldisilazane,[9] or bis(trimethylsilyl)acetamide,[10] and a catalytic amount of sodium in the presence of hexamethylphosphoric triamide. Two authors[11] who used the method of House[12] (no experimental details supplied) note that their product always contained about 30% of hexamethyldisiloxane, which could not be separated by fractional distillation.

1. Pfizer Central Research, Sandwich, Kent CT13 9NJ, England.
2. Brownbridge, P. *Synthesis* **1983**, 1, 85.
3. Reetz, M. T. *Angew. Chem., Intern. Ed. Engl.* **1982**, *21*, 96.
4. Murata, S.; Suzuki, M.; Noyori, R. *J. Am. Chem. Soc.* **1980**, *102*, 3248, and references cited therein.
5. Cazeau, P.; Moulines, F.; Laporte, O.; Duboudin, F. *J. Organomet. Chem.* **1980**, *201*, C9.
6. Emde, H.; Gotz, A.; Hofmann, K.; Simchen, G. *Justus Liebigs Ann. Chem.* **1981**, 1643.
7. Litvinova, O. V.; Baukov, Yu. I.; Lutsenko, I. F. *Dokl. Akad. Nauk SSSR* **1967**, *173*,578 ; *Chem. Abstr.* **1967**, *67*, 32720j; Lutsenko, I. F.; Baukov, Yu. I.; Dudukina, O. V.; Kramarova, E. N. *J. Organomet. Chem.* **1968**, *11*, 35.

8. Novice, M. H.; Seikaly, H. R.; Seiz, A. D.; Tidwell, T. T. *J. Am. Chem. Soc.* **1980**, *102*, 5835; Hendewerk, M. L.; Weil, D. A.; Stone, T. L.; Ellenberger, M. R.; Farneth, W. E.; Dixon, D. A. *J. Am. Chem. Soc.* **1982**, *104*, 1794.
9. Gerval, P.; Frainnet, E. *J. Organomet. Chem.* **1978**, *153*, 137.
10. Dedier, J.; Gerval, P.; Frainnet, E. *J. Organomet. Chem.* **1980**, *185*, 183.
11. Larson, G. L.; Hernandez, A. *J. Org. Chem.* **1973**, *38*, 3935.
12. House, H. O.; Czuba, L. J.; Gall, M.; Olmstead, H. D. *J. Org. Chem.* **1969**, *34*, 2324.

Appendix
Chemical Abstracts Nomenclature (Collective Index Number); (Registry Number)

Acetone trimethylsilyl enol ether: Silane, (isopropenyloxy)trimethyl- (8); Silane, trimethyl[(1-methylethenyl)oxy]- (9); (1833-53-0)

Acetone (8); 2-Propanone (9); (67-64-1)

Triethylamine (8); Ethanamine, N,N-diethyl- (9); (121-44-8)

Chlorotrimethylsilane: Silane, chlorotrimethyl- (8,9); (75-77-4)

3-HYDROXY-3-METHYL-1-PHENYL-1-BUTANONE BY CROSSED ALDOL REACTION

(1-Butanone, 3-hydroxy-3-methyl-1-phenyl-)

$$\underset{C_6H_5\overset{\overset{OSi(CH_3)_3}{|}}{C}=CH_2}{} + (CH_3)_2C=O \quad \xrightarrow{TiCl_4} \quad C_6H_5\overset{\overset{O}{\|}}{C}\text{-}CH_2\text{-}\overset{\overset{OH}{|}}{C}(CH_3)_2$$

Submitted by Teruaki Mukaiyama and Koichi Narasaka.[1]
Checked by Kathleen Hug and Clayton H. Heathcock.

1. Procedure

A 500-mL, three-necked flask is fitted with a stirring bar, rubber stopper, 100-mL pressure-equalizing dropping funnel and a three-way stop-cock which is equipped with a balloon of argon gas (Note 1). The flask is charged with 140 mL of dry methylene chloride (Note 2) and cooled in an ice bath. Titanium tetrachloride (11.0 mL) (Note 3) is added by a syringe with stirring by a magnetic stirrer, and a solution of 6.5 g of acetone in 30 mL of methylene chloride is added dropwise over a 5-min period. On completion of this addition a solution of 19.2 g of 1-phenyl-1-trimethylsiloxyethylene (Note 4) in 15 mL of methylene chloride (Note 5) is added dropwise over a 10-min period, and the mixture is stirred for 15 min.

The reaction mixture is poured into 200 mL of ice water with vigorous stirring and the organic layer is separated. The aqueous layer is extracted with two 30-mL portions of methylene chloride. The combined methylene chloride extracts are washed with two 60-mL portions of a 1:1 mixture of saturated aqueous sodium bicarbonate and water, and then with brine. The methylene chloride solution is dried over sodium sulfate and the methylene chloride is removed using a rotary evaporator.

The residue is dissolved in 30 mL of benzene, and the solution is transferred to a chromatographic column (50 mm diameter) consisting of 600 mL of silica gel. The product is eluted sequentially with (A) 1 L of 4:1 (v/v) hexane:ethyl acetate, (B) 1.5 L of 2:1 (v/v) hexane:ethyl acetate (flash chromatography) (Note 6).

The initial ca. 900 mL of the eluent is discarded. Concentration of the later fractions (ca. 1.3 L) under reduced pressure yields an oil of the pure product (Note 7). The total yield is 12.2-12.8 g (70-74%).

2. Notes

1. All the apparatus should be oven-dried before use.
2. Methylene chloride was purified by distillation from phosphorus pentoxide, then from calcium hydride, and was stored over Molecular Sieves 4A.
3. Freshly distilled titanium tetrachloride (bp 152-153°C) was used. The checkers distilled the titanium tetrachloride from calcium hydride.
4. The silyl enol ether may be obtained from the Fluka Chemical Corp., 255 Oser Avenue, Hauppauge, NY 11788. Alternatively, it may be prepared by the following modification of the procedure of Walshe and co-workers.[2] The Walshe procedure is followed exactly with 36 g (0.30 mol) of acetophenone, 41.4 g (0.41 mol) of triethylamine, 43.2 g (0.40 mol) of chlorotrimethylsilane, 60 g (0.40 mol) of sodium iodide, and 350 mL of acetonitrile. After extraction, the organic layer is dried over potassium carbonate and then concentrated with a rotary evaporator under reduced pressure. The crude product is a mixture of 97% of the desired silyl enol ether and 3% of acetophenone, as shown by gas chromatography. The crude product is distilled in a Claisen flask at a pressure of about 40 mm. After a small forerun (ca. 3

g), 52.3 g (91%) of silyl enol ether, bp 124-125.5°C, is obtained. The purity of this material is approximately 98%, as judged by gas chromatography and ^{1}H NMR spectroscopy.

5. Submitters report using 60 mL of hexane.

6. 3-Hydroxy-3-methyl-1-phenyl-1-butanone is too unstable to be purified by distillation, and is decomposed to acetophenone and acetone.

7. The initial fractions are sometimes contaminated with a less polar by-product. These fractions are condensed and purified again by column chromatography using 6:1 (v/v) hexane:ethyl acetate and then 2:1 (v/v) hexane:ethyl acetate as developing solvents. The NMR spectrum (CDCl$_3$) shows singlets at δ 1.33 (6 H, C$\underline{H}_3$), 3.12 (2 H, C$\underline{H}_2$), 4.12 (broad, OH) and complex signals between 7.24-8.01 (5 H).

3. Discussion

This procedure illustrates a general method for the preparation of crossed aldols. The aldol reaction between various silyl enol ethers and carbonyl compounds proceeds smoothly according to the same procedure (see Table I). Silyl enol ethers react with aldehydes at -78°C, and with ketones near 0°C.[3] Note that the aldol reaction of silyl enol ethers with ketones affords good yields of crossed aldols which are generally difficult to prepare using lithium or boron enolates. Lewis acids such as tin tetrachloride and boron trifluoride etherate also promote the reaction; however, titanium tetrachloride is generally the most effective catalyst.

Ketene alkyl silyl acetals may also be used as nucleophiles for the formation of β-hydroxy esters.[4] The present reaction can be carried out equally well on large or small (mmole) scales. For small scale applications, it is convenient to prepare a stock solution of titanium tetrachloride in methylene chloride. (A rubber stopper is gradually destroyed by titanium tetrachloride; therefore, a Teflon stopper should be used.) Titanium tetrachloride also promotes the aldol-type reaction between silyl enol ethers and acetals to give β-alkoxy carbonyl compounds.[5]

1. Department of Chemistry, Faculty of Science, The University of Tokyo, Hongo, Bunkyo-Ku, Tokyo 113, Japan.
2. Walshe, N. D. A., Goodwin, G. B. T.; Smith, G. C.; Woodward, F. E. *Org. Synth.* **1986**, *65*, 1.
3. Mukaiyama, T.; Banno, K.; Narasaka, K. *J. Am. Chem. Soc.* **1974**, *96*, 7503.
4. Saigo, K; Osaki, M.; Mukaiyama, T. *Chem. Lett.* **1975**, 989.
5. Mukaiyama, T.; Hayashi, M. *Chem. Lett.* **1974**, 15.

Table I. Preparation of Crossed Aldols

$$Me_3SiO-CR^1=CR^2 + R^3COR^4 \xrightarrow{TiCl_4} R^1CO-CR^2-C(OH)R^3R^4$$

	Substituents				
R^1	R^2	R^3	R^4		Yield of Aldols, %
-(CH$_2$)$_4$-		Me_2CH	H		92
		$PhCH_2$	$PhCH_2$		64
-(CH$_2$)$_3$-		$PhCH_2$	H		95
Ph	H	Me_2CH	H		94
Ph	Me	Me	H		92
Ph	Me	PhCO	H		83
Ph	Me	Me	$(CH_2)_2CO_2Me$		53

Appendix

Chemical Abstracts Nomenclature (Collective Index Number); (Registry Number)

3-Hydroxy-3-methyl-1-phenyl-1-butanone: 1-Butanone, 3-hydroxy-3-methyl-1-phenyl- (9); (43108-74-3)

Titanium tetrachloride: Titanium chloride (8,9); (7550-45-0)

Acetone (8); 2-Propanone (9); (67-64-1)

1-Phenyl-1-trimethylsiloxyethylene: Silane, trimethyl[(1-phenylvinyl)oxy]- (8); Silane, trimethyl[(1-phenylethenyl)oxy]- (9); (13735-81-4)

Acetophenone (8); Ethanone, 1-phenyl- (9); (98-86-2)

Triethylamine (8); Ethanamine, N,N-diethyl- (9); (121-44-8)

Chlorotrimethylsilane: Silane, chlorotrimethyl- (8,9); (75-77-4)

Sodium iodide (8,9); (7681-82-5)

Acetonitrile (8,9); (75-05-8)

3-DIMETHYL-1,5-DIPHENYLPENTANE-1,5-DIONE

(1,5-Pentanedione, 3,3-dimethyl-1,5-diphenyl-)

A. $\text{C}_6\text{H}_5\text{COCH}_2\text{C(OH)(CH}_3)_2 \xrightarrow[\text{(C}_2\text{H}_5)_3\text{N}]{\text{(CF}_3\text{C)}_2\text{O}} \text{C}_6\text{H}_5\text{C(O)-CH=C(CH}_3)_2$

B. $\text{C}_6\text{H}_5\text{C(O)-CH=C(CH}_3)_2 + \text{H}_2\text{C=C(OSi(CH}_3)_3)\text{-C}_6\text{H}_5 \xrightarrow{\text{TiCl}_4}$ (CH₃)₂C(CH₂C(O)C₆H₅)₂

Submitted by Koichi Narasaka.[1]
Checked by David E. Uehling and Clayton H. Heathcock.

1. Procedure

A. Isopropylideneacetophenone. A 1-L, three-necked flask is fitted with a 100-mL pressure-equalizing dropping funnel, a mechanical stirrer and a condenser which is equipped with a two-way stopcock leading to a balloon of argon gas. To the flask is added a solution of 18.4 g of 3-hydroxy-3-methyl-1-phenyl-1-butanone (Note 1) in 60 mL of dry methylene chloride. The flask is cooled in an ice bath and 28.5 g of triethylamine, a catalytic amount of 4-(N,N-dimethylamino)pyridine and 20 mL of methylene chloride are added. A solution of 25.8 g of trifluoroacetic anhydride (Note 2) in 40 mL of methylene chloride is added dropwise over a period of 15 min, and the mixture is stirred for 2.5 hr.

The ice bath is removed and the mixture is stirred for 21 hr at room temperature (about 30°C). Under vigorous stirring, 100 mL of saturated aqueous sodium carbonate, 100 mL of water and 300 mL of ether are added to the mixture. The organic layer is separated and the water layer is extracted with 100 mL of ether. The combined ether extracts are washed with brine and dried over magnesium sulfate. The ether solution is condensed using a rotary evaporator and the residue is distilled under reduced pressure to give 14.5-16.0 g (88-97% yield) of isopropylideneacetophenone (Note 3).

B. *3,3-Dimethyl-1,5-diphenylpentane-1,5-dione*. A 500-mL, three-necked flask is fitted with a mechanical stirrer, rubber septum and a two-way stopcock which is equipped with a balloon of argon gas (Note 4). To the flask is added 100 mL of dry methylene chloride, and the flask is cooled in a dry ice-acetone bath. Titanium tetrachloride (7.7 mL) (Note 5) is added by syringe through the septum. The septum is removed and replaced with a 100-mL pressure-equalizing dropping funnel containing a solution of 11.2 g of isopropylideneacetophenone in 30 mL of methylene chloride. This solution is added over a 3-min period, and the mixture is stirred for 4 min. A solution of 13.5 g of the silyl enol ether of acetophenone (Note 1) in 40 mL of methylene chloride is added dropwise with vigorous stirring over a 4-min period, and the mixture is stirred for 7 min. The reaction mixture is poured into a solution of 22 g of sodium carbonate in 160 mL of water with vigorous magnetic stirring (Note 6). The resulting white precipitate is removed by filtration through a Celite pad and the precipitate is washed with methylene chloride.

The organic layer of the filtrate is separated and the aqueous layer is extracted with two 40-mL portions of methylene chloride. The combined organic extracts are washed with 60 mL of brine and dried over sodium sulfate.

The methylene chloride solution is concentrated with a rotary evaporator and the residue is passed through a short column of silica gel (Baker 200 mesh, 400 mL) using 1.5 L of a 9:1 (v/v) mixture of hexane and ethyl acetate (Note 7). The eluent is condensed and distilled; the first fraction (bp 81-85°C/0.6 mm, 2.04 g) is a mixture of isopropylideneacetophenone and acetophenone; the second fraction (bp 85-172°C/0.6 mm, 0.42 g) is a mixture of the above substances and the desired product; the third fraction (bp 172-178°C/0.6 mm) gives 14.0-15.2 g (72-78%) of 3,3-dimethyl-1,5-diphenylpentane-1,5-dione (Note 8).

2. Notes

1. See *Organic Syntheses*, this volume p. 6.

2. Attempted dehydration using an acid catalyst or iodine failed, giving mainly acetophenone. When acetic anhydride is employed instead of trifluoroacetic anhydride, the reaction proceeds very slowly. Dehydration with excess methanesulfonyl chloride and triethylamine gives the product in high yield; however, the distilled product has a strong odor of sulfur compound.

3. The physical properties are as follows: bp 73-75°C/0.4 mm; the NMR spectrum (CCl_4) shows singlets at δ 1.93 (3 H) and 2.13 (3 H) and multiplets at 6.63 (1 H), 7.16-7.48 (3 H) and 7.71-7.91 (2 H).

4. All the apparatus should be well dried before use.

5. Freshly distilled titanium tetrachloride (bp 136.4°C) is used.

6. Stirring should be continued until the organic and aqueous layers show no acidity.

7. The submitters used Wako gel C-200.

8. The physical properties are as follows: Anal. Calcd. for $C_{19}H_{20}O_2$: C, 81.39; H, 7.19. Found: C, 81.34; H, 7.16. The ^{1}H NMR spectrum (CDCl$_3$) shows singlets at δ 1.22 (6 H, C$\underline{H}_3$) and 3.26 (4 H, C$\underline{H}_2$), and multiplet signals between 7.17-8.03 (10 H, aromatic C$\underline{H}$).

3. Discussion

The preparation of 3,3-dimethyl-1,5-diphenylpentane-1,5-dione has also been achieved from 3,3-dimethylglutaric acid and phenyllithium.[2]

The present method gives 3,3-dimethyl-1,5-diphenylpentane-1,5-dione in better yield, and is widely applicable to the preparation of various 1,5-diketones.[3] In addition, when silyl enol ethers of esters are employed instead of those of ketones, δ-keto esters can be obtained.[4]

1. Department of Chemistry, Faculty of Science, The University of Tokyo, Hongo Bunkyo-Ku, Tokyo 113, Japan.
2. Zimmerman, H. E.; Pincock, J. A. *J. Am. Chem. Soc.* **1973**, *95*, 3246.
3. Narasaka, K.; Soai, K.; Mukaiyama, T. *Chem. Lett.* **1974**, 1223; Narasaka, K.; Soai, K.; Aikawa, Y.; Mukaiyama, T. *Bull. Chem. Soc. Jpn.* **1976**, *49*, 779.
4. Saigo, K.; Osaki, M.; Mukaiyama, T. *Chem. Lett.* **1976**, 163.

Appendix

Chemical Abstracts Nomenclature (Collective Index Number);

(Registry Number)

3,3-Dimethyl-1,5-diphenylpentane-1,5-dione: 1,5-Pentanedione, 3,3-dimethyl-1,5-diphenyl- (9); (42052-44-8)

Isopropylideneacetophenone: 2-Buten-1-one, 3-methyl-1-phenyl- (9); (5650-07-7)

3-Hydroxy-3-methyl-1-phenyl-1-butanone: 1-Butanone, 3-hydroxy-3-methyl-1-phenyl- (9); (43108-74-3)

Triethylamine (8); Ethanamine, N,N-diethyl- (9); (121-44-8)

4-(N,N-Dimethylamino)pyridine: Pyridine, 4-(dimethylamino)- (8); 4-Pyridinamine, N,N-dimethyl- (9); (1122-58-3)

Trifluoroacetic anhydride: Acetic acid, trifluoro-, anhydride (8,9); (407-25-0)

Titanium tetrachloride: Titanium chloride (8,9); (7550-45-0)

Acetophenone silyl enol ether: Silane, trimethyl[(1-phenylvinyl)oxy]- (8); Silane, trimethyl[(1-phenylethenyl)oxy]- (9); (13735-81-4)

RING EXPANSION AND CLEAVAGE OF SUCCINOIN DERIVATIVES: SPIRO[4.5]DECANE-1,4-DIONE AND ETHYL 4-CYCLOHEXYL-4-OXOBUTANOATE

(Cyclohexanebutanoic acid, γ-oxo-, ethyl ester)

Submitted by Eiichi Nakamura and Isao Kuwajima.[1]
Checked by Jens Wolff and Ian Fleming.

1. Procedure

A. *Spiro[4.5]decane-1,4-dione* (2). In a dry, 200-mL, two-necked flask with one neck connected to a nitrogen source to maintain a positive pressure and with the other covered with a rubber septum is placed a magnetic stirring bar. Boron trifluoride etherate (5.04 mL, 40.0 mmol) (Note 1) and 40 mL of dry methylene chloride (Note 2) are introduced with a hypodermic syringe and the solution is cooled to ca. -75°C with a dry ice/hexane bath. A mixture of cyclohexanone diethyl ketal (6.88 g, 40 mmol) (Note 3) and 1,2-bis(trimethyl-silyloxy)cyclobut-1-ene (9.20 g, 40 mmol) (Note 4) in 20 mL of dry methylene chloride is added during 10 min. The resulting yellow solution is stirred for

30 min at that temperature and 8 mL of trifluoroacetic acid (Note 5) is added. The mixture is warmed to room temperature and stirred for 2 hr (Note 6) before addition of 40 mL of water. The mixture is extracted three times with 100 mL-portions of ether, and the combined extract is washed successively with a 30-mL portion of water, saturated aqueous sodium bicarbonate (2 x 45 mL), and 30 mL of saturated sodium chloride. The extract is dried over anhydrous magnesium sulfate, filtered, and concentrated on a rotary evaporator at aspirator pressure. The crude, viscous oily material is distilled at 0.05 mm with a Kugelrohr apparatus (Note 7) with an oven temperature at 75-80°C to obtain 5.40-6.00 g (81-90%) of spiro[4.5]decane-1,4-dione, which crystallizes on cooling to room temperature, mp 61-62°C (Note 8).

B. *Ethyl 4-cyclohexyl-4-oxobutanoate* (**4**). In a dry, 100-mL, two-necked flask with one neck connected to a nitrogen source to maintain a positive pressure and with the other covered with a rubber septum is placed a magnetic stirring bar. Tin tetrachloride (7.50 g, 28.8 mmol) (Note 9) and 15 mL of dry methylene chloride (Note 2) are introduced with a hypodermic syringe and the solution is cooled to ca. -75°C with a dry ice/hexane bath. A mixture of cyclohexanone diethyl ketal (4.82 g, 28.0 mmol) (Note 3) and 1,2-bis(trimethylsilyloxy)cyclobut-1-ene (6.44 g, 28.0 mmol) (Note 4) in 10 mL of methylene chloride is added during 10 min. The yellow solution is stirred for 15 min at -75°C, and for 15 min at -30°C, during which period the solution turns heterogeneous (Note 10). Water (20 mL) and ether (50 mL) are added and the organic layer is separated. The aqueous layer is extracted twice with ether (50 mL) and the combined organic layers are washed successively with 3 x 10-mL portions of 1 N hydrochloric acid, and 20-mL portions each of water, aqueous sodium bicarbonate, and saturated sodium chloride. The oily product (6.30 g) obtained after drying (anhydrous magnesium sulfate) and concentration on a rotary evaporator is distilled to give an analytically pure keto ester (**4**) (5.27-5.44 g, 90-93%) as a fraction boiling at 110-112°C, 2.5 mm, or 84°C, 0.2 mm (Note 11).

2. Notes

1. Boron trifluoride etherate (Hashimoto Kasei Chemical, Osaka) was distilled before use.

2. Methylene chloride was distilled from phosphorus oxide and stored over molecular sieves.

3. Cyclohexanone diethyl ketal was prepared according to a procedure by Howard and Lorette; see *Org. Synth., Collect. Vol. V* **1973**, 292; bp 80-83°C, 18 mm. The checkers prepared it by keeping cyclohexanone (50 g), triethyl orthoformate (75 g) and concentrated hydrochloric acid (0.2 mL) in absolute ethanol (30 mL) for 10 hr at room temperature, followed by treatment with sodium hydroxide until the solution is basic.

4. 1,2-Bis(trimethylsilyloxy)cyclobut-1-ene was prepared in ca. 80% yield on a 0.5-mol scale by Method 2 described by Bloomfield and Nelke: see *Org. Synth.* **1977**, *57*, 1.

5. Commercially available trifluoroacetic acid (Tokyo Kasei Co.) was used as received.

6. If trifluoroacetic acid treatment is omitted, the aldol-type adduct, 2-(1-ethoxycyclohexyl)-2-(trimethylsilyloxy)cyclobutanone (**1**), is obtained in high (ca. 90%) yield; bp 85-90°C (bath temp), 0.05 mm; IR (neat) cm^{-1}: 1789 (s, C=O); ^{1}H NMR (CCl$_4$) δ: 0.11 (s, 9 H), 0.9-2.2 (m, including t, J = 7 at δ 1.10), 2.2-2.9 (m, 3 H), 3.2-3.7 (AB part of ABX$_3$, 2 H). Treatment of purified **1** with trifluoroacetic acid gives **2** in nearly quantitative yield.

7. Kugelrohr distillation ovens are manufactured by Büchi Glasapparatefabrik.

8. The checkers found variable amounts (0-15%) of the ester (**4**) in the product. This could easily be removed by recrystallization of the diketone from light petroleum (bp 40-60°C) to give needles, mp 62-64°C. The product has the following spectral properties: IR (CCl_4) cm^{-1}: 1720 (vs); ^{1}H NMR (CCl_4) δ: 1.58 (br s, 10 H), 2.65 (s, 4 H); MS (70 eV) m/e (relative intensity) 166.0983 (M^+, 100, calcd for $C_{10}H_{14}O_2$; 166.0994), 137 (24), 124 (36), 112 (87), 111 (83), 109 (27), 85 (38), 81 (48), 67 (95), 56 (48), 54 (45), 53 (42), 41 (58), 30 (64).

9. Tin tetrachloride (Yoneyama Yakuhin Co.) was distilled before use.

10. If the mixture is quenched with triethylamine before aqueous workup, the intermediate enol silyl ether **3** is obtained; bp 110-115°C (bath temp), 0.04 mm; IR (neat) cm^{-1}: 1745 (s), 1680 (m); ^{1}H NMR (CCl_4) δ: 0.12 (s, 9 H), 1.3-2.6 (m, 14 H, including t, J = 7, at 1.26 and s at 2.36), 4.10 (q, J = 7). Addition of aldehydes, acetals, or phenylsulfenyl chloride at this stage gives the respective aldol and sulfenylated products.

11. The product has the following spectral properties: IR (CCl_4), cm^{-1}: 1739 (s), 1713 (s); ^{1}H NMR (CCl_4) δ: 0.8-1.9 (m, 18 H, including t, J = 1.28, $CH_2\underline{CH_3}$), 4.15 (q, J = 7, $OC\underline{H_2}CH_3$); MS (70 eV) m/e (relative intensity) 212 (M^+, 5), 167 (20), 129 (55), 111 (28), 101 (80), 83 (100), 55 (72), 41 (37), 29 (34).

3. Discussion

The present reactions are based on the novel rearrangement of succinoin derivatives such as **1** which are obtainable in high yield by the reaction of

1,2-bis(trimethylsilyloxy)cyclobut-1-ene with carbonyl compounds. The first procedure, Part A, illustrates a general method for preparing a wide range of spiro[4.n]alkane-1,4-diones as well as useful 2-mono- and 2,2-disubstituted cyclopentane-1,3-diones (Table I).[2] The combination of an aldol reaction[3] and a skeletal rearrangement provides a highly efficient new approach to these synthetically interesting molecules.[4] The reaction can be performed either in a single pot or as a two-stage operation by isolating the initial aldol adduct **1**. The merit of the sequence as a spiro annelation method is illustrated by the synthesis of a 5,8-methanospiro[4.5]decanedione from norcamphor (Table I). γ-Keto acids and cyclopent-2-ene-1,4-diones also become available from ketals in a few steps.

The second procedure, Part B, illustrates an easy synthesis of γ-keto esters by "reductive succinoylation" of a ketal function.[5] It is useful not only for the preparation of keto esters, but also as a four-carbon chain-elongation reaction starting from ketones. The reaction is applicable to a diverse range of ketals as shown in Table II. The enol silyl ether intermediate **3** can either be isolated or used in situ for further elaboration. Fluoride-[6] and Lewis-acid catalyzed aldol reactions cleanly give aldol adducts,[6] and the reaction with phenylsulfenyl chloride gives α-phenylthio ketones in high yield.[5]

3 →[E⁺] [cyclohexyl with E and C(=O)CH₂CH₂COOEt substituents]

Cyclobutanone 1 is also useful for the stereoselective construction of quaternary carbon centers[7] and 2,3-substituted cyclopentenones.[8] The synthetic utility of the chemistry of 1 and related compounds has been reviewed.[7,9]

1. Department of Chemistry, Tokyo Institute of Technology, Meguro, Tokyo 152, Japan.
2. Nakamura, E.; Kuwajima, I. *J. Am. Chem. Soc.* **1977**, *99*, 961.
3. Mukaiyama, T.; Banno, K.; Narasaka, K. *J. Am. Chem. Soc.* **1974**, *96*, 7503.
4. Cf., Hengartner, U.; Chu, V. *Org. Synth.* **1978**, *58*, 83.
5. Nakamura, E.; Hashimoto, K.; Kuwajima, I. *J. Org. Chem.* **1977**, *42*, 4166.
6. Nakamura, E.; Shimizu, M.; Kuwajima, I.; Sakata, J.; Yokoyama, K.; Noyori, R. *J. Org. Chem.* **1983**, *48*, 932.
7. (a) Shimada, J.-i.; Hashimoto, K.; Kim, B. H.; Nakamura, E.; Kuwajima, I. *J. Am. Chem. Soc.* **1984**, *106*, 1759; Kuwajima, I. In "Current Trends in Organic Synthesis", Nozaki, H., Ed.; Pergamon Press: Oxford, 1983; pp. 311-322.
8. Nakamura, E.; Shimada, J.-i.; Kuwajima, I. *J. Chem. Soc., Chem. Commun.* **1983**, 498.
9. Bloomfield, J. J.; Owsley, D. C.; Nelke, J. M. *Org. React.* **1976**, *23*, 259.

Table I. Synthesis of Cyclopentane-1,3-diones

Acetal or Ketal	Product	Yield (aldol x rearrangement)[a]
benzaldehyde diethyl acetal		94 x 97%
decanal dimethyl acetal		90 x 87%
3-pentanone dimethyl ketal		92 x 87%
cyclododecanone dimethyl ketal		92 x 94% (91%)[b]
2-norbornanone dimethyl acetal		60 x 92%

[a]The results in this table were obtained by the two stage procedure (i.e. the isolation of the initial adduct, e.g., 1, is involved). [b]Yield from the single-pot procedure as described in the text.

Table II. Reductive Succinoylation

Ketal	Product	Yield
3-pentanone dimethyl ketal	(Et)₂CH-C(O)-CH₂CH₂-COOMe	87
4-tert-butylcyclohexanone dimethyl ketal	4-tert-butylcyclohexyl-C(O)-CH₂CH₂-COOMe	92
2-norbornanone dimethyl ketal	2-norbornyl-C(O)-CH₂CH₂-COOMe	90
cyclododecanone dimethyl ketal	cyclododecyl-C(O)-CH₂CH₂-COOMe	91
adamantanone dimethyl ketal	2-adamantyl-C(O)-CH₂CH₂-COOMe	68

Appendix

Chemical Abstracts Nomenclature (Collective Index Number);

(Registry Number)

Spiro[4.5]decane-1,4-dione (9); (39984-92-4)

Ethyl 4-cyclohexyl-4-oxobutanoate: Cyclohexanebutanoic acid, γ-oxo-, ethyl ester (9); (54966-52-8)

Boron trifluoride etherate: Ethyl ether, compd. with boron fluoride (BF_3) (1:1) (8); Ethane, 1,1'-oxybis-, compd. with trifluoroborane (1:1) (9); (109-63-7)

Cyclohexanone diethyl ketal (8); Cyclohexane, 1,1-diethoxy- (9); (1670-47-9)

1,2-Bis(trimethylsilyloxy)cyclobut-1-ene: Silane, (1-cyclobuten-1,2-ylenedioxy)bis[trimethyl- (8); Silane, [1-cyclobutene-1,2-diylbis(oxy)]bis[trimethyl- (10); (17082-61-0)

Trifluoroacetic acid: Acetic acid, trifluoro- (8,9); (76-05-1)

Tin tetrachloride: Tin chloride (8); Stannane, tetrachloro- (9); (7646-78-8)

THE STETTER REACTION:

3-METHYL-2-PENTYL-2-CYCLOPENTEN-1-ONE (DIHYDROJASMONE)

(2-Cyclopenten-1-one, 3-methyl-2-pentyl-)

A. $CH_3(CH_2)_5CHO \; + \; CH_2=CH-\overset{O}{\overset{\|}{C}}-CH_3 \quad \xrightarrow[\text{ethanol, 80°C}]{\text{HO}\diagdown\text{—}\overset{CH_2C_6H_5}{\overset{|}{N^+}}\diagdown S \; Cl^- \atop NEt_3,} \quad CH_3(CH_2)_5\overset{O}{\overset{\|}{C}}CH_2CH_2\overset{O}{\overset{\|}{C}}CH_3$

B. $CH_3(CH_2)_5\overset{O}{\overset{\|}{C}}CH_2CH_2\overset{O}{\overset{\|}{C}}CH_3 \quad \xrightarrow[-H_2O]{OH^-} \quad$ 3-methyl-2-pentyl-2-cyclopenten-1-one

Submitted by H. Stetter, H. Kuhlmann and W. Haese.[1]
Checked by Rodney A. Badger and James D. White.

1. Procedure

A. 2,5-Undecanedione. A 1000-mL, three-necked, round-bottomed flask equipped with a mechanical stirrer, short gas inlet tube, and an efficient reflux condenser fitted with a potassium hydroxide drying tube is charged with 26.8 g (0.1 mol) of 3-benzyl-5-(2-hydroxyethyl)-4-methyl-1,3-thiazolium chloride (Note 1), 500 mL of absolute ethanol, 77.2 g (1.1 mol) of 3-buten-2-one (Note 2), 60.6 g (0.6 mol) of triethylamine (Note 3), and 114.2 g (1.0 mol) of heptanal (Note 4). A slow stream of nitrogen (Note 5) is started and the mixture is stirred and heated in an oil bath at 80°C. After 16 hr the reaction mixture is cooled to room temperature and concentrated by rotary evaporation. Then 500 mL of chloroform is added to the residue and the mixture is washed with 200 mL of dilute hydrochloric acid (5%), 200 mL of saturated sodium hydrogen carbonate solution and, finally, with two 200-mL portions of water. After the solution is dried with anhydrous magnesium sulfate, the chloroform is distilled off and the residue is fractionated under reduced pressure through a 30-cm Vigreux column. The main fraction is collected at 80-82°C/0.3 mm. The yield is 130-138 g (71-75% based on heptanal) of a colorless distillate, which solidifies on standing at room temperature, mp 33-34°C (Note 6, 7).

B. *3-Methyl-2-pentyl-2-cyclopenten-1-one (Dihydrojasmone).* 2,5-Undecanedione (92.1 g, 0.5 mol) is added to a solution of 16.0 g (0.4 mol) of sodium hydroxide in 800 mL of water and 200 mL of ethanol in a 2000-mL round-bottomed flask. The mixture is refluxed for 6 hr, cooled to room temperature, and extracted with ether. The combined ether phases are dried with magnesium sulfate, and the solution is separated from the drying agent and concentrated at room temperature under reduced pressure. The residual oil is distilled through a 30-cm Vigreux column. The pure compound boils at 65-67°C/0.5 mm and weighs 70-73 g (84-88% based on the diketone) (Note 8).

2. Notes

1. 3-Benzyl-5-(2-hydroxyethyl)-4-methyl-1,3-thiazolium chloride is supplied by Fluka AG, Buchs, Switzerland and by Tridom Chemical, Inc., Hauppague, New York. The thiazolium salt may also be prepared as described in *Org. Synth.* **1984**, *62*, 170.

2. 3-Buten-2-one was used as obtained from Fluka AG, Buchs, Switzerland.

3. Triethylamine was dried with potassium hydroxide and distilled. Instead of triethylamine, sodium acetate (32.8 g, 0.4 mol), which has been dried under vacuum at 100°C for 1 day, can also be used.

4. Heptanal was supplied by Aldrich Chemical Company, Inc. It was freshly distilled before use.

5. The nitrogen flow rate should be one bubble per second.

6. A boiling point of 141°C/14 mm and a melting point of 33°C is recorded.[2] The diketone exhibits the following spectral characteristics: IR (CDCl$_3$) cm^{-1}: 1710; ^{1}H NMR (CDCl$_3$) δ: 0.77-1.67 (m, 11 H, C$\underline{H}_2$ and C$\underline{H}_3$); 2.13 (s, 3 H, C-CH$_3$); 2.30-2.60 (m, 2 H, C$\underline{H}_2$); 2.67 (s, 4 H, COC$\underline{H}_2$C$\underline{H}_2$CO).

7. The checkers obtained a second fraction from the distillation (13.5 g, 7.4%), bp 97-105°C/0.15 mm, which solidified upon cooling. Recrystallization of this material from hexane gave a colorless solid, mp 26-27°C, which was identified from its infrared, NMR, and mass spectra as 8-hydroxy-7-tetradecanone. This product arises via a "benzoin-type" condensation, catalyzed by the thiazolium salt, of heptanal.

8. A boiling point of 122-124°C/12 mm is recorded.[2] The cyclopentenone exhibits the following spectral characteristics: IR (neat) cm^{-1}: 1695 and 1640; ^{1}H NMR (CDCl$_3$) δ: 0.77-1.50 (m, 9 H, C$\underline{H}_2$ and C$\underline{H}_3$); 2.03 (s, 3 H, C$\underline{H}_3$); 2.13-2.57 (m, 6 H, C$\underline{H}_2$). For fragrance it is advisable to destroy malodorous by-products by the method described in Note 9.

9. The use of sodium acetate instead of triethylamine (see Note 3) is an alternative and is followed by an oxidizing treatment of the diketone: 100 g of 2,5-undecanedione is dissolved in 500 mL of methylene chloride and treated with 10 g of an oxidizing reagent (Note 10). The mixture is refluxed for 3 hr, filtered and washed with three 100-mL portions of water. The organic phase is dried with sodium sulfate and distilled. This material is converted into dihydrojasmone by procedure B, and a last, efficient distillation (Fisher, slit tube-system, HMS 500) leads to chromatographic purity greater than 99 per cent.

10. Oxidizing reagent:[3] To a solution of 500 g (0.5 mol) of chromium (VI) oxide and 300 mL of water is added 250 g of silica gel (silica gel 60, E. Merck, Darmstadt, West Germany). The mixture is shaken at 30-35°C for 1 hr. After this, the water is removed on a rotary evaporator to yield a yellow-orange, free-flowing powder.

3. Discussion

2,5-Undecanedione and the cyclization to dihydrojasmone were first described by H. Hunsdiecker.[2] The natural jasmine odor components and the artificial substitutes have been the goal of many investigations.[4] Our method of preparing 2,5-undecanedione by addition of heptanal to 3-buten-2-one[5] is only one example of a wide range of reactions involving the conjugate addition to electron-deficient double bonds.[6]

A large variety of aldehydes has been used in the addition to butenone (we give some characteristic examples):

1. Simple straight-chain aliphatic aldehydes (C_2 to C_{12} tested) and mono α-branched aldehydes.[7]

2. Conjugated unsaturated aldehydes (e.g., citral, β,β-dimethylacrolein[8]).

3. Aldehydes that contain isolated double bonds, such as 10-undecenal, citronellal, 3-cyclohexene-1-carboxaldehyde and norbornene carboxaldehyde.[8,9]

4. Aldehydes containing a variety of other functional groups, e.g., ether groups,[10] the phthalimido group,[11] keto, ester and nitrile groups.[12,13]

5. Heterocyclic and aromatic aldehydes[7,12] (e.g., furan-2-carboxaldehyde, thiophen-2-carboxaldehyde, the pyridine carboxaldehydes, benzaldehyde, and diverse substituted benzaldehydes).

Variations have been made in the activated system also. Higher homologues of butenone (e.g., 1-penten-3-one, tert-butyl vinyl ketone) react in the same manner, as does phenyl vinyl ketone. The same variety of functional groups as shown before may be possible in the side chain of the ketone.[14]

Additions to acrylic esters and acrylonitrile[15] and to arylidene and alkylidene β-dicarbonyl compounds[16] are possible.

The addition of aldehydes to α,β-unsaturated sulfones yields γ-diketones.[17]

The mechanism of the thiazolium ion-catalyzed conjugate addition reactions[6] is considered to be analogous to the Lapworth mechanism for the cyanide-catalyzed benzoin condensation, the thiazolium ylide playing the role of cyanide. The resulting intermediate carbanion is presumed to be the actual Michael donor. After conjugate addition to the activated olefin, the thiazolium ylide is eliminated to form the product and regenerate the catalyst.

1. Institut für Organische Chemie der Rheinisch-Westfälischen Technischen Hochschule Aachen, West Germany.
2. Hunsdiecker, H. *Ber. Dtsch. Chem. Ges.* **1942**, *75*, 447.
3. Singh, R. P.; Subbarao, H. N.; Dev, S. *Tetrahedron* **1979**, *35*, 1789.
4. Van der Gen, A. *Parfums, Cosmet, Savons Fr.* **1972**, *2*, 356; *Chem. Abstr.* **1973**, *78*, 7735n.
5. Stetter, H.; Kuhlmann, H. *Synthesis* **1975**, 379.
6. Stetter, H. *Angew. Chem., Intern. Ed. Engl.* **1976**, *15*, 639; Stetter, H.; Kuhlmann, H. Ger. Offen. 2437219, 1974, Bayer AG, *Chem. Abstr.* **1976**, *84*, 164172t.

7. Stetter, H.; Kuhlmann, H. *Chem. Ber.* **1976**, *109*, 3426.
8. Stetter, H.; Hilboll, G.; Kuhlmann, H. *Chem. Ber.* **1979**, *112*, 84.
9. Stetter, H.; Landscheidt, A. *Chem. Ber.* **1979**, *112*, 1410 and 2419.
10. Stetter, H.; Mohrmann, K.-H.; Schlenker, W. *Chem. Ber.* **1981**, *114*, 581.
11. Stetter, H.; Lappe, P. *Chem. Ber.* **1980**, *113*, 1890.
12. Stetter, H.; Basse, W.; Kuhlmann, H.; Landscheidt, A.; Schlenker, W. *Chem. Ber.* **1977**, *110*, 1007.
13. Stetter, H.; Basse, W.; Wiemann, K. *Chem. Ber.* **1978**, *111*, 431.
14. Stetter, H.; Nienhaus, J. *Chem. Ber.* **1980**, *113*, 979.
15. Stetter, H.; Basse, W.; Nienhaus, J. *Chem. Ber.* **1980**, *113*, 690.
16. Stetter, H.; Jonas, F. *Chem. Ber.* **1981**, *114*, 564.
17. Stetter, H.; Bender, H.-J. *Chem. Ber.* **1981**, *114*. 1226.

Appendix

Chemical Abstracts Nomenclature (Collective Index Number);

(Registry Number)

3-Methyl-2-pentyl-2-cyclopenten-1-one: 2-Cyclopenten-1-one, 3-methyl-2-pentyl- (8,9); (1128-08-1)

2,5-Undecanedione (8,9); (7018-92-0)

3-Benzyl-5-(2-hydroxyethyl)-4-methyl-1,3-thiazolium chloride (8,9); (4568-71-2)

3-Buten-2-one (8,9); (78-94-4)

Triethylamine (8); Ethanamine, N,N-diethyl- (9); (121-44-8)

Heptanal (8,9); (111-71-7)

PREPARATION AND THREE-CARBON + TWO-CARBON CYCLOADDITION OF A CYCLOPROPENONE KETAL: CYCLOPROPENONE 1,3-PROPANEDIOL KETAL
(4,8-Dioxaspiro[2.5]oct-1-ene)

Submitted by Dale L. Boger, Christine E. Brotherton, and Gunda I. Georg.[1]
Checked by Steven K. Davidsen and Clayton H. Heathcock.

1. Procedure

A. *2-(Bromomethyl)-2-(chloromethyl)-1,3-dioxane.*[2] A 100-mL, round-bottomed flask is equipped with a 10-mL Dean-Stark apparatus and a condenser. The flask is charged with 30.0 g (0.138 mol) of 1-bromo-3-chloro-2,2-dimethoxypropane (Note 1), 10.0 mL (0.138 mol) of 1,3-propanediol (Note 2), and 3 drops of concentrated sulfuric acid. The resulting solution is heated (bath temperature 140°C) for 8 hr (Note 3) with distillative removal of methanol (ca. 11 mL). The mixture is allowed to cool to room temperature and the crude product is partitioned in 150 mL of pentane and 40 mL of water. The organic phase is dried with magnesium sulfate and the solvent is removed under reduced pressure (Note 4). Distillation (1 mm, 90-95°C) yields 25.5-27.7 g (81-88%) of 2-(bromomethyl)-2-(chloromethyl)-1,3-dioxane, mp 57-59°C (Note 5).

B. *Cyclopropenone 1,3-propanediol ketal*.[3] A 1000-mL, three-necked, round-bottomed flask is equipped with a gas inlet, an acetone-dry ice condenser with a drying tube containing potassium hydroxide pellets, a stopper and a magnetic stirring bar. The flask is placed in an acetone-dry ice bath and anhydrous ammonia (Note 6) is condensed into the flask (400 mL). A small piece of potassium metal (ca. 0.5 g) is added to the liquid ammonia and the cooling bath is removed. A catalytic amount of anhydrous ferric chloride (0.1 g) is added and the reaction mixture is allowed to warm to reflux temperature, at which time the deep blue color turns to gray. The remaining potassium metal (12.2 g, 0.31 g-atom total) is added in 0.5-g pieces over ca. 30 min. The reaction mixture is allowed to stir until a gray suspension results (20-30 min). A -50°C cooling bath is placed under the flask and the stopper is replaced with a 125-mL, pressure-equalized dropping funnel containing 22.9 g (0.1 mol) of 2-(bromomethyl)-2-(chloromethyl)-1,3-dioxane in 50 mL of anhydrous ether. This solution is added dropwise to the solution of freshly generated potassium amide over 15 min while the temperature is maintained at -50°C (Note 7). After the solution is stirred for 3 hr at -50°C to -60°C, solid ammonium chloride is added slowly to quench the excess potassium amide (Note 8). The cooling bath is removed and the ammonia is allowed to evaporate. During the course of the evaporation, anhydrous ether (350 mL total) is added dropwise through the addition funnel to replace the ammonia. After the temperature has reached 0°C, the brown reaction mixture is filtered by suction through a coarse fritted glass filter to remove the inorganic salts and the salts are washed twice with 25 mL of anhydrous ether. The combined ethereal filtrate and washes are concentrated under reduced pressure (80-100 mm, 30°C) to a constant weight (ca. 4-5 hr) (Note 9). The residue is transferred to a 50-mL, round-bottomed flask fitted with a water-cooled, short-path distillation head, and the product is distilled (1.25 mm; 30-35°C) into an ice-cooled receiver. Cyclopropenone 1,3-propanediol ketal is obtained as a colorless liquid (6.1-7.8 g, 55-70% yield, Note 10).

C. *5,5-Dicyano-4-phenylcyclopent-2-enone 1,3-propanediol ketal*. Benzylidenemalononitrile (Note 2) (3.85 g, 25 mmol) and cyclopropenone 1,3-propanediol ketal (5.6 g, 50 mmol) are combined in 25 mL of dry distilled toluene (Note 11) in a sealed tube (Note 12) with a magnetic stirring bar. The reaction mixture is heated at 80°C for 6.5 hr with stirring. The crude reaction mixture is filtered through a short plug of glass wool and applied to a Waters Associates Prep LC/System 500, eluting with 2.5:1 hexane/ethyl acetate (Note 13). The fractions containing product are combined and concentrated under reduced pressure to give 4-4.2 g (60-64%) of 5,5-dicyano-4-phenylcyclopent-2-enone 1,3-propanediol ketal as a white solid: mp 139-141°C (ethanol, Note 14).

2. Notes

1. The preparation[4] of this material has been described in detail: Breslow, R.; Pecoraro, J.; Sugimoto, T. *Org. Synth.* **1977**, *57*, 41.

2. The submitters employed material available from Aldrich Chemical Company, Inc. without further purification.

3. Shorter reaction times result in incomplete transketalization.

4. The checkers found that the product crystallizes, either upon removal of solvent, or in the condenser during the following distillation. To prevent the condenser tube becoming plugged by the crystalline product, an air-cooled, rather than a water-cooled condenser jacket should be used.

5. The product has the following spectral properties: ^{1}H NMR (CDCl$_3$) δ: 1.78 (m, 2 H, C$\underline{H}_2$), 3.70 (s, 2 H, C$\underline{H}_2$Cl), 3.80 (s, 2 H, C$\underline{H}_2$Br), 3.96 (t, 4 H, J = 6, OC$\underline{H}_2$); IR (CHCl$_3$) cm^{-1}: 3030, 3000, 2900, 1485, 1430, 1245, 1202, 1158, 1135, 1105 (s), 1020.

6. Commercial anhydrous ammonia is employed without further drying.

7. Crystals which may form at the tip of the addition funnel are scraped off and allowed to drop into the reaction flask.

8. The checkers added a total of 25 g of solid ammonium chloride in portions of approximately 2 g with a spatula.

9. The checkers removed the solvent with a rotary evaporator at 20 mm and 25°C; under these conditions only 1 hr is required to concentrate the solution to a constant weight.

10. This material should be stored under an argon atmosphere below 0°C. The product has the following spectral properties: ^{1}H NMR (CDCl$_3$) δ: 1.83 (m, 2 H, C$\underline{H}_2$), 4.01 (t, 4 H, J = 6, OC$\underline{H}_2$), 7.84 (s, 2 H, C$\underline{H}$=C$\underline{H}$); ^{13}C NMR (CDCl$_3$) δ: 25.4 (CH$_2$), 65.6 (OCH$_2$), 80.5 (OCO), 125.1 (C=C); IR (film) cm^{-1}: 3101, 2980, 2870, 1600, 1475, 1460, 1435, 1370, 1300, 1275, 1155, 1090, 1030, 935, 910, 865, 740.

11. Toluene was distilled from calcium hydride under a nitrogen atmosphere.

12. The resealable glass tube was fabricated from an Ace Glass medium-walled straight tube. The tube was permanently sealed on one end and the other end remained internally threaded. The tube was sealed with a solid Teflon plug fitted with a FETFE O-ring. Various sizes of such tubes are now available from Ace Glass.

13. The submitters used medium pressure chromatography[5] on 25 x 1000 mm of 230-400 mesh silica gel with a 25 x 250 mm scrubber column. The eluant (3:1:1 hexane:ethyl acetate:methylene chloride) was passed through the column at a rate of 15 mL/min, collecting 15 mL fractions. The checkers were unable to obtain pure product using a gravity column or flash chromatography.

14. The product has the following spectral properties: ^{1}H NMR (CDCl$_3$) δ: 1.71 (dt, 1 H, J = 13.8, J = 3.8, OCH$_2$C$\underline{H}_2$CH$_2$O), 2.1-2.3 (broad m, 1 H, OCH$_2$C$\underline{H}_2$CH$_2$O), 3.9-4.3 (m, 4 H, OC$\underline{H}_2$CH$_2$C$\underline{H}_2$O), 4.63 (t, 1 H, J = 2.3, allylic C$\underline{H}$), 6.25 (dd, 1 H, J = 6.3, J = 2.0, CH-CH=C$\underline{H}$-C), 6.62 (dd, 1 H, J = 6.3, J = 2.6, CH-C$\underline{H}$=CH-C), 7.3-7.5 (m, 5 H, phenyl); IR (CHCl$_3$) cm^{-1}: 3070, 2920, 2290 (C≡N), 1510, 1465, 1350, 1260, 1210, 1175, 1140, 1100, 1080, 1040, 870, 705.

3. Discussion

Although a number of multistep procedures are available for the introduction of five-membered carbocycles, their direct formation in a thermal cycloaddition is rare.[6] Interest in the potential application of such a three-carbon + two-carbon cyclopentane cycloaddition has been derived from the expectation that such a process could prove to be an effective complement to the four-carbon + two-carbon Diels-Alder reaction which is used extensively in the regio- and stereocontrolled preparation of functionalized six-membered carbocycles.

Cyclopropenone ketals, of which cyclopropenone 1,3-propanediol ketal (1) is a representative and unusually stable example, have proven to be useful equivalents of the 1,3-dipole (i) in a regiospecific three-carbon + two-carbon cycloaddition with electron-deficient olefins, (eq 1). Table I shows representative results of a study of this reaction.[7]

Studies have defined the scope of the thermal reactions of cyclopropenone ketals which are characterized by their thermal, reversible ring opening to provide reactive intermediates best represented as delocalized singlet vinyl carbenes, three-carbon 1,1-/1,3-dipoles without octet stabilization, (eq 2).

In addition to the representative [3 + 2] cycloaddition reactions shown in Table I, the delocalized singlet vinyl carbenes have been shown to participate as $\pi 2_a$ components of non-linear cheletropic $[\pi 2_s + \pi 2_a]$ cycloadditions to provided cyclopropanes with an observable endo effect,[8] and as $\pi 2_s$ components of $[\pi 4_s + \pi 2_s]$ cycloadditions with selected dienes to provide cycloheptadienes,[9] (Scheme 1). This thermal reactivity of cyclopropenone ketals

Scheme 1

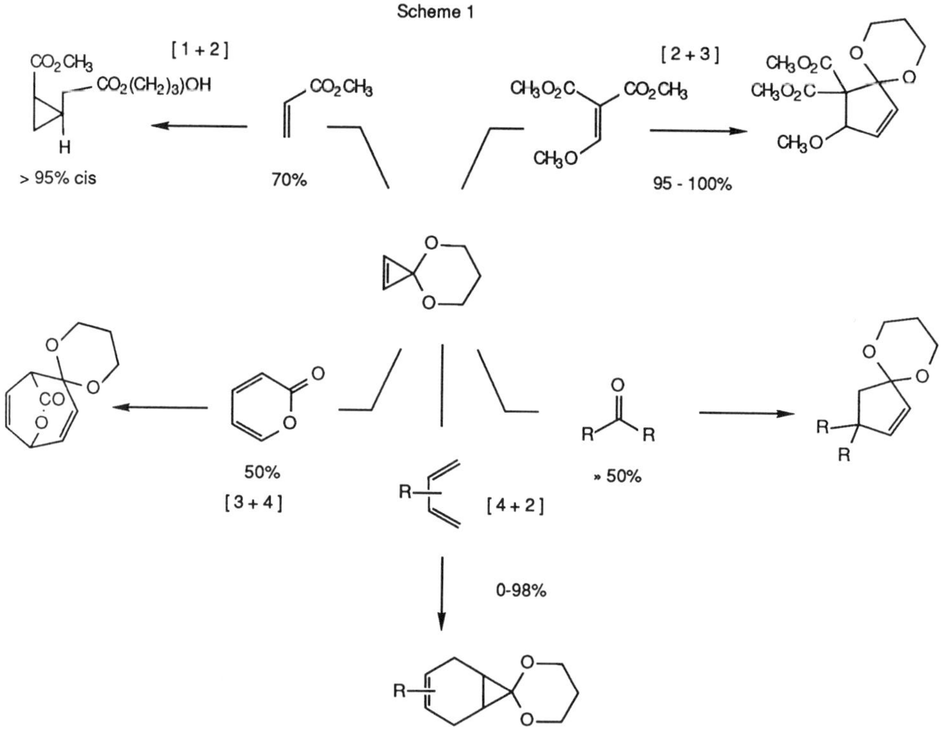

complements their dual participation as strained olefins in normal ($HOMO_{diene}$ controlled) and inverse electron demand ($LUMO_{diene}$ controlled) Diels-Alder reactions with electron-rich, electron-deficient, and neutral dienes under room temperature, thermal, and pressure-promoted Diels-Alder conditions.[10]

Table I. Reactions of Cyclopropenone Ketal 1 with Electron-deficient Olefins

Substrate	Conditions[a] equiv 1, temp. ºC(time hr)	Product	%Yield
(bicyclic lactone with CO₂CH₃)	1.0-2.0,80(12)	(cycloadduct with CO₂CH₃)	42%
(bicyclic ketone with CO₂Et)	1.0,75(13)	(cycloadduct with CO₂Et)	45%
RO₂C-C(Ph)=C-CO₂R	1.5,75(10) 2.5,75(32)	(cyclopentene with RO₂C, RO₂C, Ph)	48% 60%
EtO₂C-C(CH₃)=C-CO₂Et	1.5,75(15)	(cyclopentene with EtO₂C, EtO₂C, CH₃)	57%
CH₃O₂C-C(OCH₃)=C-CO₂CH₃	2.0,80(5)	(cyclopentene with CH₃O₂C, CH₃O₂C, CH₃O)	84%
NC-C(Ph)=C-CN	2.0,80(4)	(cyclopentene with NC, NC, Ph)	60-64%

(a) All reactions were run in benzene (0.5 - 2.0 M in substrate) under nitrogen unless otherwise noted.

1. Department of Medicinal Chemistry, The University of Kansas, Lawrence, KS 66045. Present address: Department of Chemistry, Purdue University, West Lafayette, IN 47907.
2. Butler, G. B.; Herring, K. H.; Lewis, P. L.; Sharpe, V. V., III; Veazey, R. L. *J. Org. Chem.* **1977**, *42*, 679.
3. Albert, R. M.; Butler, G. B *J. Org. Chem.* **1977**, *42*, 674.
4. Baucom, K. B.; Butler, G. B. *J. Org. Chem.* **1972**, *37*, 1730.
5. Meyers, A. I.; Slade, J.; Smith, R. K.; Mihelich, E. D.; Hershenson, F. M.; Liang, C. D. *J. Org. Chem.* **1979**, *44*, 2247.
6. Trost, B. M.; Chan, D. M. T. *J. Am. Chem. Soc.* **1983**, *105*, 2315, 2326; Danheiser, R. L.; Carini, D. J.; Basak, A. *J. Am. Chem. Soc.* **1981**, *103*, 1604; Mayr, H.; Seitz, B.; Halberstadt-Kausch, I. K. *J. Org. Chem.* **1981**, *46*, 1041; Little, R. D.; Muller, G. W.; Venegas, M. G.; Carroll, G. L.; Bukhari, A.; Patton, L.; Stone, K. *Tetrahedron* **1981**, *37*, 4371; Noyori, R. *Acc. Chem. Res.* **1979**, *12*, 61.
7. Boger, D. L.; Brotherton, C. E. *J. Am. Chem. Soc.* **1984**, *106*, 805.
8. Boger, D. L.; Brotherton, C. E. *Tetrahedorn Lett.* **1984**, *25*, 5611.
9. Boger, D. L.; Brotherton, C. E. *J. Org. Chem.* **1985**, *50*, 3425.
10. Boger, D. L.; Brotherton, C. E. *Tetrahedron*, in press.

Appendix

Chemical Abstracts Nomenclature (Collective Index Number);

(Registry Number)

Cyclopropenone 1,3-propanediol ketal: 4,8-Dioxaspiro[2.5]oct-1-ene (9); (60935-21-9)

2-(Bromomethyl)-2-(chloromethyl)-1,3-dioxane: 1,3-Dioxane, 2-(bromomethyl)-2-(chloromethyl)- (9); (60935-30-0)

1-Bromo-3-chloro-2,2-dimethoxypropane: 2-Propanone, 1-bromo-3-chloro-, dimethyl acetal (8); Propane, 1-bromo-3-chloro-2,2-dimethoxy- (9); (22089-54-9)

1,3-Propanediol (8,9); (504-63-2)

5,5-Dicyano-4-phenylcyclopent-2-enone 1,3-propanediol ketal: 6,10-Dioxaspiro[4.5]dec-3-ene-1,1-dicarbonitrile, 2-phenyl- (11); (88442-12-0)

Benzylidenemalononitrile: Malononitrile, benzylidene- (8); Propanedinitrile, (phenylmethylene)- (9): (2700-22-3)

1,2,3,4,5-PENTAMETHYLCYCLOPENTADIENE

(1,3-Cyclopentadiene, 1,2,3,4,5-pentamethyl-)

A. $\text{CH}_3\text{CH}=\text{C}(\text{Br})(\text{CH}_3)$ + 2 Li $\xrightarrow{\text{Et}_2\text{O}}$ $\text{CH}_3\text{CH}=\text{C}(\text{Li})(\text{CH}_3)$ + LiBr

$\text{CH}_3\text{CH}=\text{C}(\text{Li})(\text{CH}_3)$ + ethyl acetate $\xrightarrow{\text{Et}_2\text{O}}$ $\xrightarrow{\text{H}_2\text{O}}$ 3,4,5-trimethyl-2,5-heptadien-4-ol

B. 3,4,5-trimethyl-2,5-heptadien-4-ol $\xrightarrow[-\text{H}_2\text{O}]{\text{C}_7\text{H}_7\text{SO}_3\text{H}}$ 1,2,3,4,5-pentamethylcyclopentadiene

Submitted by Richard S. Threlkel,[1] John E. Bercaw,[1] Paul F. Seidler,[2] Jeffrey M. Stryker,[2] and Robert G. Bergman.[2]

Checked by David E. Hill and James D. White.

1. Procedure

A. 3,4,5-Trimethyl-2,5-heptadien-4-ol. Lithium wire (1/8") is cut up into approximately 1-cm lengths and washed with hexane (Note 1). A mixture of the cut-up lithium (21 g, 3.0 mol) in 100 mL of diethyl ether from a freshly opened can is stirred well under argon in a 2-L, three-necked, round-bottomed flask equipped with a reflux condenser and a 250-mL addition funnel. 2-Bromo-2-butene (cis and trans mixture) is purified and dried by passing it through a 2 x 15-cm column of basic alumina I (Note 2). An addition funnel is charged with 2-bromo-2-butene (200 g, 1.48 mol) and a small amount of the alkene is added dropwise to the flask until reaction begins, as shown by warming of the reaction mixture and formation of bubbles on the surface of the lithium (Note 3). At this point the mixture is diluted with an additional 900 mL of fresh diethyl ether, and the remainder of the 2-bromo-2-butene is added at a rate sufficient to maintain gentle reflux. Stirring is continued for 1 hr following completion of this addition, after which ethyl acetate (66 g, 0.75 mol) diluted with an equal volume of fresh diethyl ether is added dropwise via the addition funnel. The reaction mixture turns from yellow-orange to milky-yellow with this addition. It is then poured into 2 L of saturated aqueous ammonium chloride. The ethereal layer is separated, and the aqueous layer is adjusted to approximately pH 9 with hydrochloric acid. The aqueous layer is extracted three times with diethyl ether. The combined ethereal layers are dried over magnesium sulfate, filtered, and concentrated to 100-200 mL by rotary evaporation.

B. *1,2,3,4,5-Pentamethylcyclopentadiene.* A mixture of 13 g (0.068 mol) of p-toluenesulfonic acid monohydrate and 300 mL of diethyl ether is stirred under argon in a 1-L, three-necked, round-bottomed flask equipped with a reflux condenser and a 250-mL addition funnel. The concentrate from above is added as quickly as possible to the flask from the addition funnel, maintaining a gentle reflux. As the reaction proceeds, a water layer separates. The mixture is stirred for 1 hr after the addition is completed and then washed with saturated aqueous sodium bicarbonate until the washings remain basic. The ethereal layer is separated, and the combined aqueous layers are extracted three times with diethyl ether. The combined ethereal layers are dried over sodium sulfate. Diethyl ether is removed by rotary evaporation, and the crude product is distilled under reduced pressure (bp 55-60°C, 13 mm): yield 73-75 g (73-75%) (Note 4).

2. Notes

1. High sodium lithium (1-2% sodium) is preferred in order to facilitate initiation of the reaction. (The checkers used 1% sodium in lithium wire purchased from Lithium Corporation of America, Bessemer City, NC).

2. While published procedures have used samples of 2-bromo-2-butene as obtained from the supplier without further purification, impurities in some batches often make it difficult, if not impossible, to start the reaction safely.

3. *Caution! It is imperative to add only a few mL of the 2-bromo-2-butene and to patiently wait for the reaction with the lithium to begin. Addition of too much 2-bromo-2-butene too soon may lead to a violent reaction.*

4. Pentamethylcyclopentadiene has the following spectral properties; IR (neat) cm^{-1}: 2940, 2890, 2840, 2730, 1670, 1655, 1440, 1370; ^{1}H NMR (CDCl$_3$) δ: 1.00 (3 H, d, J = 7.6, C$\underline{H}_3$CH), 1.8 (12 H, br s, C$\underline{H}_3$C=), 2.45 (1 H, q, J = 6.5, C$\underline{H}$CH$_3$).

3. Discussion

The procedure described here is a modification of that previously published.[3] Specifically, it is frequently insufficient to use the 2-bromo-2-butene as obtained from the supplier without purification using basic alumina I. Such purification assures facile reaction with lithium. Furthermore, the large volumes of diethyl ether used in the past are unnecessary and may inhibit initiation of the reaction of 2-bromo-2-butene with lithium. Finally, while dry solvents and reagents are required, diethyl ether from a freshly opened can is sufficiently free of water, and distillation from lithium aluminum hydride is unnecessary.

1,2,3,4,5-Pentamethylcyclopentadiene is a useful aromatic building block for the preparation of other compounds. It can be converted to many salts of its conjugate base with alkali metals or strong bases such as butyllithium.[4] These pentamethylcyclopentadienyl anion salts as well as the diene itself can be transformed into η^5-pentamethylcyclopentadienyl ligands of organotransition metal complexes by many known methods.[4]

1. Division of Chemistry and Chemical Engineering, California Institute of Technology, Pasadena, CA 91125.
2. Department of Chemistry, University of California, Berkeley, CA 94720.
3. Threlkel, R. S.; Bercaw, J. E. *J. Organomet. Chem.* **1977**, *136*, 1.
4. Feitler, D.; Whitesides, G. M. *Inorg. Chem.* **1976**, *15*, 466; Green, M. L. H.; Pardy, R. B. A. *J. Chem. Soc., Dalton Trans.* **1979**, 355 and references therein.

Appendix

Chemical Abstracts Nomenclature (Collective Index Number);

(Registry Number)

1,2,3,4,5-Pentamethylcyclopentadiene: 1,3-Cyclopentadiene, 1,2,3,4,5-pentamethyl- (9); (4045-44-7)

3,4,5-Trimethyl-2,5-heptdien-4-ol: 2,5-Heptadien-4-ol, 3,4,5-trimethyl- (9); (64417-15-8)

Lithium (8,9); (7439-93-2)

Diethyl ether: Ethyl ether (8); Ethane, 1,1'-oxybis- (9); (60-29-7)

2-Bromo-2-butene (cis and trans mixture): 2-Butene, 2-bromo- (9); (13294-71-8)

p-Toluenesulfonic acid monohydrate (8); Benzenesulfonic acid, 4-methyl-, monohydrate (9); (6192-52-5)

ALKOXYCARBONYLATION OF PROPARGYL CHLORIDE:

METHYL 4-CHLORO-2-BUTYNOATE

(2-Butynoic acid, 4-chloro-, methyl ester)

$$ClCH_2C{\equiv}CH \quad \xrightarrow[\text{2. } ClCO_2CH_3]{\text{1. } CH_3Li} \quad ClCH_2C{\equiv}CCO_2CH_3$$

Submitted by M. Olomucki and J. Y. Le Gall.[1]
Checked by H. T. M. Le and M. F. Semmelhack.

1. Procedure

Caution! Propargyl chloride, methyl chloroformate and methyl 4-chloro-2-butynoate are vesicants and lachrymators. This preparation should be conducted in a ventilated hood and protective gloves should be worn.

A 250-mL, round-bottomed flask is equipped with stirring bar (Note 1), thermometer, and a pressure-equalizing dropping funnel (dried in an oven at 80°C, assembled while still hot) and the system is placed under argon (Note 2). Via syringe, 7.45 g (7.16 mL, 0.1 mol) of propargyl chloride (Note 3) and 35 mL of anhydrous diethyl ether (Note 4) are added. The solution is stirred and cooled to -50°C to -60°C (Note 5) with an alcohol-dry ice bath (Note 6). Under gentle argon pressure, 72.4 mL of 1.41 M solution of methyllithium in diethyl ether (Note 7) is added dropwise over ca. 20 min (Note 8). Stirring is continued for 15 min and 18.9 g (0.2 mol) (Note 9) of methyl chloroformate (Note 10) is added through the dropping funnel over ca. 10 min. The reaction mixture is allowed to warm slowly (3-4 hr) to 0°C to -5°C during which time a fine precipitate appears. Water (40 mL) is added dropwise with efficient stirring; the ether layer is separated and the aqueous layer is extracted two times with ether. The combined ether solutions are dried over anhydrous magnesium sulfate, and the ether is removed under reduced pressure with a rotary evaporator. The residual liquid is distilled in a simple distillation assembly under reduced pressure, affording 10.7-11.1 g (81-83%) of methyl 4-chloro-2-butynoate as a colorless liquid, bp 41°C (0.25 mm), n_D^{22} 1.4728 (Note 11).

2. Notes

1. The submitters specify a mechanical stirrer; the checkers find magnetic stirring to be more convenient and equally effective.

2. The system was alternately evacuated with an oil pump and then filled with argon three or more times, and a positive pressure was maintained throughout the reaction period. The submitters used nitrogen in place of argon.

3. Propargyl chloride (98% purity), obtained from Fluka AG, was used without further purification.

4. Ether was dried over sodium wire.

5. Although cooling to -20 to -30°C is sufficient, the addition of methyllithium is more convenient at lower temperatures.

6. The checkers used a constant temperature refrigerated bath (Cryocool).

7. Methyllithium in the form of solutions in diethyl ether is supplied by Aldrich Chemical Company, Inc. in rubber septum stoppered bottles, which should be stored in a refrigerator.

8. The solution of methyllithium was conveniently handled using techniques for the manipulation of air-sensitive reagents.[2]

9. Lower yields were obtained when less methyl chloroformate was used. Thus, the yield was about 55% when the reaction was performed with one equivalent of methyl chloroformate, and 70-72% with 1.5 mol of the latter per mol of propargyl chloride.

10. Methyl chloroformate was used as supplied by Fluka AG (98% purity).

11. The product gives satisfactory elemental analysis and shows the following IR spectrum (film) cm^{-1}: 2267, 1725, 720.

3. Discussion

Methyl 4-chloro-2-butynoate has been prepared[3] in 54% yield by treatment of 4-chloro-2-butynoic acid with 10% sulfuric acid in methanol. 4-Chloro-2-butynoic (chlorotetrolic) acid has been prepared[3] in 40% yield by chromic acid oxidation of 4-chloro-2-butyn-1-ol (the latter obtained[4] in 45% yield by the reaction of 2-butyne-1,4-diol with thionyl chloride) or in 85% yield by treatment of the lithium derivative of propargyl chloride with carbon dioxide.[5]

The present synthesis illustrates a convenient preparation of chlorotetrolic esters which can be performed in one step starting from commercially available and inexpensive products; it is faster and gives better yields as compared with the overall yields of the multistep preparations described earlier. Since chlorotetrolic acid is not an intermediate in this synthesis, the necessity of distilling this explosive product is eliminated. In contrast to the acid, explosions were never observed during distillations of the lower boiling chlorotetrolic esters.

Other 4-chloro-2-butynoic esters can be obtained by varying the alkyl chloroformates. Thus, ethyl 4-chloro-2-butynoate was prepared[6] in the same way in 60% yield, and tert-butyl 4-chloro-2-butynoate in 73% yield; the procedure could probably be further generalized. When butyllithium is used in these syntheses instead of methyllithium, much lower (ca. 30%) yields are obtained.

Chlorotetrolic esters are small, highly functionalized, reactive molecules; of particular interest is the possibility of using them as reagents for chemical modification of biological macromolecules. Different protein nucleophiles react under mild conditions with methyl 4-chloro-2-butynoate by addition across the triple bond and/or substitution of chlorine[7] while the triple bond and the ester group are involved in the reaction of chlorotetrolic esters with nucleic acid bases.[8]

1. Laboratoire de Biochimie Cellulaire, College de France, 75005 Paris, France. This work was supported in part by a grant N° 80.7.0296 from the Ministry of Industry and Research.

2. Kramer, G. W.; Levy, A. B.; Midland, M. M. In H. C. Brown, "Organic Synthesis via Boranes", Wiley: New York, 1975; p. 191.

3. Olomucki, M. *Compt. Rend. Hebd. Seances Acad. Sci.* **1958**, *246*, 1877-1879; Olomucki, M. *Ann. Chim. (Paris)* **1960**, *5*, 845-904.

4. Dupont, G.; Dulou, R.; Lefebvre, G. *Bull. Soc. Chim. Fr.* **1954**, 816-820.

5. Battioni, J.-P.; Chodkiewicz, W. *C. R. Hebd. Seances Acad. Sci., Ser. C.* **1966**, *263*, 761-763; Battioni, J.-P. *Bull. Soc. Chim. Fr.* **1969**, 911-914.

6. Olomucki, M.; Le Gall, J.-Y.; Barrand, I. *J. Chem. Soc., Chem. Commun.* **1982**, 1290-1291.

7. Diopoh, J.; Olomucki, M. *Bioorg. Chem.* **1982**, *11*, 463-477.

8. Olomucki, M.; Le Gall, J.-Y.; Colinart, S.; Durant, F.; Norberg, B.; Evrard, G. *Tetrahedron Lett.* **1984**, *25*, 3471-3474; Olomucki, M.; Le Gall, J.-Y.; Roques, P.; Blois, F.; Colinart, S. *Nucleosides and Nucleotides* **1985**, *4*, 161-163.

Appendix
Chemical Abstracts Nomenclature (Collective Index Number); (Registry Number)

Propargyl chloride: Propyne, 3-chloro- (8); 1-Propyne, 3-chloro- (9); (624-65-7)

Methyl 4-chloro-2-butynoate: 2-Butynoic acid, 4-chloro-, methyl ester (9); (41658-12-2)

Methyl chloroformate: Formic acid, chloro-, methyl ester (8); Carbonochloridic acid, methyl ester (9); (79-22-1)

Methyllithium: Lithium, methyl- (8,9); (917-54-4)

1,4-BIS(TRIMETHYLSILYL)BUTA-1,3-DIYNE

(Silane, 1,3-butadiyne-1,4-diylbis[trimethyl-)

$$(CH_3)_3 Si-C\equiv CH \xrightarrow[(CH_3)_2NCH_2CH_2N(CH_3)_2]{O_2,\ CuCl} (CH_3)_3 Si-C\equiv C-C\equiv C-Si(CH_3)_3$$

Submitted by Graham E. Jones, David A. Kendrick, and Andrew B. Holmes.[1]
Checked by James Armstrong and Clayton H. Heathcock.

1. Procedure

A. Copper(I) chloride - tetramethylethylenediamine complex. A 200-mL, three-necked, round-bottomed flask equipped with a magnetic stirring bar, rubber septum, nitrogen inlet tube, and bubbler is charged with acetone (90 mL) and copper(I) chloride (5 g, 51 mmol) (Note 1). After the flask is purged with nitrogen, the mixture is stirred and N,N,N',N'-tetramethylethylenediamine (TMEDA) (2.5 mL, 16.6 mmol) (Note 2) is added. Stirring is maintained for 30 min, and the solid material is then allowed to settle, leaving a clear deep blue-green solution of the CuCl-TMEDA catalyst which is used in the oxidative coupling reaction.

B. 1,4-Bis(trimethylsilyl)buta-1,3-diyne (BTMSBD). A 1-L, four-necked flask, equipped with a mechanical stirrer (Note 3), dry ice cold-finger condenser (Note 4), sintered gas inlet, and a swan-neck adapter which supports a thermometer and rubber septum is charged with acetone (300 mL) and trimethylsilylacetylene (50 g, 0.51 mmol) (Note 5). The reaction mixture is agitated and a rapid stream of oxygen is passed through the solution (Note 6). The supernatant solution containing the CuCl-TMEDA catalyst is transferred by syringe in 5-mL portions into the reaction vessel. The temperature rises as the catalyst is added and should have reached 35°C after about 75% of the catalyst has been added. When this temperature is reached, external ice cooling is applied to moderate the exothermic reaction (Note 7). The remaining catalyst is added and the temperature is maintained in the range of 25-30°C for 2.5 hr (Note 8). When the reaction is complete there should be no evidence of trimethylsilylacetylene condensing on the cold trap. Agitation and oxygenation are then stopped.

The acetone is removed by evaporation with a rotary evaporator, and the residue is dissolved in petroleum ether (bp 30-40°C, 150 mL) (Note 9) and shaken in a separatory funnel with 3 M aqueous hydrochloric acid (150 mL). The phases are separated and the aqueous phase is washed with petroleum ether (bp 30-40°C, 3 x 150 mL). The combined organic layers are washed with saturated aqueous sodium chloride (50 mL), dried (Na_2SO_4), and evaporated to dryness with a rotary evaporator. The solid residue is dissolved in hot methanol (400 mL) to which has been added 3 M aqueous hydrochloric acid (4 mL). The solution may be filtered at this stage if it is necessary to remove colored insoluble impurities. Water is then added dropwise until recrystallized material is permanently present. The solution is allowed to cool, finally in ice, and crystalline bis(trimethylsilyl)butadiyne (BTMSBD) is collected. The material is washed with a small portion of ice-cold methanol-water (50:50 v/v; 50 mL), and dried in the air to give bis(trimethylsilyl)-butadiyne (31-35 g, 68-76%), mp 111-112°C (lit.[2,3] 107-108°C) (Note 10). A further 3-5 g (6-10%) of the product is obtained from the mother liquors (Notes 11 and 12).

2. Notes

1. Copper(I) chloride, S.L.R. grade, was used as supplied by Fisons Scientific Apparatus Ltd. Best results are obtained with fresh samples of copper(I) chloride which usually contains $\leq$ 2% copper(II) chloride. Further purification[4] did not improve the yield of BTMSBD. The checkers used Mallinckrodt Chemical Company, Analytical Reagent Grade copper(I) chloride, without further purification.

2. N,N,N',N'-Tetramethylethylenediamine (TMEDA) (98%) was used as supplied by B.D.H. The checkers used 99% TMEDA, as supplied by Aldrich Chemical Company, Inc.

3. Mechanical stirring is adequate for a 50-g scale, as described herein, and magnetic stirring is sufficient for a 5-g scale. The submitters ran the procedure on a 200-g scale, and found that use of a Vibro-mixer is essential to obtain satisfactory oxygenation of the reaction mixture. The Vibro-mixer model E-1 was supplied by Chemap AG, Alte Landstr. 415, CH-8708, Mannedorf, Switzerland.

4. The reactant is sufficiently volatile in the fast oxygen stream that substantial loss of material occurs unless a cold finger condenser with a large contact area, charged with dry ice-2-propanol, is used. The use of a short Vigreux column between the dry ice condenser and the reaction vessel is strongly recommended to provide additional protection against loss of material as an aerosol.

5. Trimethylsilylacetylene is prepared by silylation of ethynylmagnesium chloride as described in the accompanying procedure in Organic Syntheses. It is also commercially available from the Aldrich Chemical Company, Inc.

6. *CAUTION*. Although no hazard has been encountered in this reaction, due care should be taken with acetylenic compounds in an atmosphere of oxygen. The experiment should be conducted in a well-ventilated hood behind a safety shield and away from any source of ignition. Dilution of exit gases (T-joint) with nitrogen is strongly advised.

7. On two occasions the temperature was observed to reach 35°C well before 75% of the catalyst had been added by the checkers. The temperature should be monitored closely during this addition. If ice cooling is necessary, it is important to lower the temperature only to 25°C; otherwise the reaction will become too sluggish. After the internal temperature is brought to 25°C, further cooling is not needed during addition of the remaining catalyst.

8. A deep blue-green coloration should be evident throughout the addition of the catalyst. The color is determined by the rate of oxygen flow. Too high a flow rate can lead to over-oxidation, producing a black precipitate, whereas too low a flow rate can lead to over-reduction of the catalyst, with the green color fading to be replaced by an orange-red precipitate. Both factors reduce the yield of BTMSBD.

9. The checkers used pentane instead of petroleum ether.

10. The checkers observed mp 109-110°C for all crops.

11. Yields of 95% on a 5-g scale, and 70-75% on a 200-g scale were reproducibly obtained by the submitters.

12. Pure bis(trimethylsilyl)butadiyne exhibits the following spectroscopic data: IR (CCl_4) cm^{-1}: 2080 (s), 1250 (s), and 650 (s); UV (C_6H_{12}) nm max(ε): 224 (80), 235 (150), 248 (260), 262 (345), and 278 (250); ^{1}H NMR $(CDCl_3$, 250 MHz) δ: 0.22 (18 H).

3. Discussion

Oxidative coupling of terminal acetylenes in the presence of copper(I) catalysts is the best method of preparing symmetrically substituted butadiyne derivatives,[5] and has been applied to the coupling of trimethylsilylacetylene.[6] Better yields are obtained using the Hay procedure in which the catalyst is the TMEDA complex of copper(I) chloride.[7] The procedure submitted here is an improved version of Walton and Waugh's synthesis of BTMSBD by the Hay coupling of trimethylsilylacetylene.[2] BTMSBD has also been prepared by silylation of butadiynedimagnesium bromide[3] and chloride[8] in moderate yield, and more recently from the dilithium derivative in good yield.[25]

BTMSBD is a very convenient source of butadiyne, an extremely useful, but dangerously explosive chemical.[9] It is also a synthon for the vinylacetylene anion. A single trimethylsilyl group can selectively be replaced by reaction with electrophiles (Friedel-Crafts reaction) to give trimethylsilylbutadiynyl ketones.[2]

$$(CH_3)_3Si-C\equiv C-C\equiv C-Si(CH_3)_3 \ + \ RCOCl \ \xrightarrow{AlCl_3} \ RCO-C\equiv C-C\equiv C-Si(CH_3)_3$$

Alternatively, a more nucleophilic anionic reagent can be generated by selective cleavage of a single trimethylsilyl group with methyllithium-lithium bromide complex.[10] This lithiobutadiyne derivative will react with electrophiles such as carbonyl compounds[10,11] or primary alkyl iodides.[12]

$$(CH_3)_3Si-C\equiv C-C\equiv C-Si(CH_3)_3$$

$$\downarrow CH_3Li/LiBr$$

$$LiC\equiv C-C\equiv C-Si(CH_3)_3$$

RI ↙   ↘ R'R"CO

$$R-C\equiv C-C\equiv C-Si(CH_3)_3 \qquad R'R"C(OH)C\equiv C-C\equiv C-Si(CH_3)_3$$

Regio- and stereoselective reduction of the non-silylated triple bond, either by partial catalytic hydrogenation,[13,14,15] or by lithium aluminum hydride reduction of the propargylic alcohols,[11,16,17] afford (after desilylation), respectively, terminal (Z)- and (E)-enynes. Furthermore, the remaining trimethylsilyl group in both silylated diynes and enynes may be replaced by another electrophile in a second Friedel-Crafts reaction.[18]

Such reactions have found a variety of applications in natural products synthesis.[11,13,14,16,17,19]

BTMSBD reacts with a variety of nucleophiles to give novel heterocycles such as selenophen,[20] tellurophen,[21] and pyrazoles.[22] It has also been used in [2+4] cycloaddition/cycloreversion sequences to prepare ethynyl-substituted pyridazines[23] and furans.[24]

1. University Chemical Laboratory, Lensfield Road, Cambridge, CB2 1EW, U.K.
2. Walton, D. R. M.; Waugh, F. *J. Organomet. Chem.* **1972**, *37*, 45.
3. Shikhiev, I. A.; Shostakovskiĭ, M. F.; Kayutenko, L. A. *Dokl. Akad. Nauk Azerbaĭdzhan, S.S.R.* **1959**, *15*, 21; *Chem. Abstr.* **1959**, *53*, 15957e.
4. Keller, R. N.; Wycoff, H. D. *Inorg. Synth.* **1946**, *2*, 1.
5. Eglinton, G.; McCrae, W. In "Advances in Organic Chemistry: Methods and Results", Raphael, R. A.; Taylor, E. C.; Wynberg, H., Eds.; Wiley: New York, 1963; Vol. 4, p. 225; Cadiot, P.; Chodkiewicz, W. In "Chemistry of Acetylenes", Viehe, H. G., Ed.; Marcel Dekker: New York, 1969, Chapter 9.
6. Minh, L. Q.; Billiotte, J.-C.; Cadiot, P. *C.R. Hebd. Seances Acad. Sci.* **1960**, *251*, 730.
7. Hay, A. S. *J. Org. Chem.* **1962**, *27*, 3320.
8. Ballard, D. H.; Gilman, H. *J. Organomet. Chem.* **1968**, *15*, 321.
9. Armitage, J. B.; Jones, E. R. H.; Whiting, M. C. *J. Chem. Soc.* **1951**, 44.
10. Holmes, A. B.; Jennings-White, C. L. D.; Schulthess, A. H.; Akinde, B.; Walton, D. R. M. *J. Chem. Soc., Chem. Commun.* **1979**, 840.
11. Salaün, J.; Ollivier, J. *Nouv. J. Chim.* **1981**, *5*, 587.
12. Holmes, A. B.; Jones, G. E. *Tetrahedron Lett.* **1980**, *21*, 3111.
13. Holmes, A. B.; Raphael, R. A.; Wellard, N. K. *Tetrahedron Lett.* **1976**, 1539.

14. Kobayashi, A.; Shibata, Y.; Yamashita, K. *Agric. Biol. Chem.* **1975**, *39*, 911.
15. Shakhovskoi, B. G.; Stadnichuk, M. D.; Petrov, A. A. *Zh. Obshch. Khim.* **1964**, *34*, 2625; *J. Gen. Chem. U.S.S.R. (Engl. Transl.)* **1964**, *34*, 2646.
16. Patrick, T. B.; Melm, G. F. *J. Org. Chem.* **1979**, *44*, 645.
17. Holmes, A. B.; Jennings-White, C. L. D.; Kendrick, D. A. *J. Chem. Soc., Chem. Commun.* **1984**, 1594.
18. Jones, G. E.; Holmes, A. B. *Tetrahedron Lett.* **1982**, *23*, 3203.
19. Holmes, A. B.; Jennings-White, C. L. D.; Kendrick D. A. *J. Chem. Soc., Chem. Commun.* **1983**, 415.
20. Jacobs, P. M.; Davis, M. A.; Norton, H. *J. Heterocycl. Chem.* **1977**, *14*, 1115.
21. Lohner, W.; Praefcke, K. *Chem. Ber.* **1978**, *111*, 3745.
22. Birkofer, L.; Richtzenhain, K. *Chem. Ber.* **1979**, *112*, 2829.
23. Birkofer, L.; Hänsel, E. *Chem. Ber.* **1981**, *114*, 3154.
24. Liotta, D.; Saindane, M.; Ott, W. *Tetrahedron Lett.* **1983**, *24*, 2473.
25. Zweifel, G.; Rajagopalan, S. *J. Am. Chem. Soc.* **1985**, *107*, 700.

Appendix

Chemical Abstracts Nomenclature (Collective Index Number);

(Registry Number)

1,4-Bis(trimethylsilyl)buta-1,3-diyne: 2,7-Disilaocta-3,5-diyne, 2,2,7,7-tetramethyl- (8); Silane, 1,3-butadiyne-1,4-diylbis[trimethyl- (9); (4526-07-2)

Copper(I) chloride: Copper chloride (8,9); (7758-89-6)

N,N,N',N'-Tetramethylenediamine: Ethylenediamine, N,N,N',N'-tetramethyl- (8); 1,2-Ethanediamine, N,N,N',N'-tetramethyl- (9); (110-18-9)

Trimethylsilylacetylene: Silane, ethynyltrimethyl- (8,9); (1066-54-2)

TRIMETHYLSILYLACETYLENE

(Silane, ethynyltrimethyl-)

A. $CH_3CH_2CH_2CH_2Cl \xrightarrow[\text{Tetrahydrofuran}]{\text{Mg}} CH_3CH_2CH_2CH_2MgCl$

B. $CH_3CH_2CH_2CH_2MgCl \xrightarrow{HC\equiv CH} HC\equiv C-MgCl$

C. $HC\equiv C-MgCl + (CH_3)_3SiCl \longrightarrow HC\equiv C-Si(CH_3)_3$

Submitted by Andrew B. Holmes and Chris N. Sporikou.[1]
Checked by Dallas D. Crotts, Bruce E. Eaton, and Clayton H. Heathcock.

1. Procedure

A. Butylmagnesium chloride. A dry, 1-L, three-necked, round-bottomed flask is equipped with a sealed mechanical stirrer (Note 1), a 250-mL pressure-equalizing dropping funnel, and a reflux condenser to the top of which is attached a T-piece connected at one end to a supply of dry nitrogen, and at the other to an oil or mercury bubbler. At the same time a dry, 2-L, three-necked, round-bottomed flask, fitted with a sealed mechanical stirrer and two swan neck adapters, is prepared, with one adapter holding a gas inlet and a calcium chloride drying tube, and the other supporting a thermometer and a 1-L pressure-equalizing dropping funnel.

The 1-L flask is charged with magnesium turnings (39.6 g, 1.65 g-atom) (Note 2) and dry tetrahydrofuran (THF) (150 mL), the mixture is heated to reflux temperature under an atmosphere of dry nitrogen, and a crystal of iodine is added. The dropping funnel is filled with 1-chlorobutane (173 mL, 152.5 g, 1.65 mol) (Note 3) and a portion (15 mL) is added to the boiling THF mixture. The source of heat is removed. After the reaction has commenced, the THF begins to boil more vigorously and a further volume (400 mL) of THF is added to the reaction mixture. Then the remainder of the chlorobutane is added slowly at a rate sufficient to maintain the reaction under reflux. Finally, the reaction mixture is stirred and heated under reflux until all the magnesium has been consumed (0.5-1 hr).

B. *Ethynylmagnesium chloride.* While the butylmagnesium chloride is being prepared, the 2-L flask is filled with dry THF (500 mL), which is saturated by bubbling acetylene through it for 0.5-1 hr (Note 4). The warm (ca. 60°C) butylmagnesium chloride is rapidly poured under a liberal blanket of nitrogen into the 1-L dropping funnel which is then flushed with nitrogen before being stoppered. The rapid flow of acetylene is maintained. The 2-L flask is cooled to -5°C in a dry ice-acetone bath, the acetylene is bubbled rapidly through the THF (Note 4), and the butylmagnesium chloride is added dropwise to the stirred reaction mixture at a rate sufficient to maintain the temperature below 20°C (Note 5). This addition requires 1 hr; then acetylene is bubbled through the reaction mixture for a further 0.5 hr (the mixture cools to about 5°C during this period). The acetylene supply is disconnected and replaced by dry nitrogen.

C. *Trimethylsilylacetylene.* A solution of chlorotrimethylsilane (152 mL, 130 g, 1.197 mol) (Note 6) in dry THF (100 mL) is placed in the 1-L dropping funnel and is added (20 min) to the cooled and stirred solution of ethynylmagnesium chloride at a rate sufficient to maintain a reaction temperature of about 15-20°C (Note 7). Finally, the dropping funnel is replaced by an efficient double surface condenser and calcium chloride drying tube, and the reaction mixture is heated under reflux for 1 hr (Note 8). The reflux condenser is replaced by a distillation head and a double surface condenser is connected to a receiver flask which is cooled in an ice bath (Note 9). The reaction mixture is distilled under nitrogen with stirring until all the azeotrope of trimethylsilylacetylene and THF (700-800 mL, bp ca. 66°C) has distilled (Notes 10, 11). The distillate is washed with ice-water portions (10 x 500 mL) to remove the THF. Washing is continued until the organic layer stays constant in volume (Note 12). Distillation (Note 10) of the organic layer under an atmosphere of nitrogen through a short Vigreux column (Note 13) gives trimethylsilylacetylene, bp 50-52°C/760 mm, n_D^{20} 1.391 (lit.[2,3,4] 53.5°C/762 mm, n_D^{21} 1.3900; 52°C/760 mm, n_D^{20} 1.3900; 52°C/760 mm, n_D^{20} 1.3935) in yields ranging from 72.5 g (62%) to 87.5 g (75%) (Notes 14, 15).

2. Notes

1. The checkers used an efficient magnetic stirrer.

2. The magnesium turnings for Grignard reactions were supplied by Fisons Scientific Apparatus, Loughborough. If less than 1.65 g-atom of magnesium is used, the final product will be contaminated with 1-chlorobutane.

3. It is essential that the 1-chlorobutane be free of 1-butanol. 1-Chlorobutane (Aldrich Chemical Company, Inc.) was purified by rapid passage through basic alumina (activity 1) before use.

4. Gaseous acetylene is introduced at the rate of about 20 L/hr and is purified by passage through a cold trap (-78°C), followed by bubbling through concentrated sulfuric acid and finally passage over sodium hydroxide pellets. These operations must be conducted in a well-ventilated fume hood. The checkers found that an insufficient acetylene flow rate in this and the next step results in the formation of butyltrimethylsilane and bis(trimethylsilyl)acetylene.

5. The temperature reaches 15°C after 0.5 hr, and 20°C after 0.8 hr. Ethynylmagnesium halides can rapidly disproportionate to bis(chloromagnesium)-acetylene and acetylene at higher temperatures.[2] It is important to maintain the reaction mixture at or below 20°C and to have an excess of acetylene in order to prevent formation of the bis(magnesium chloride). The checkers found that the ethynylmagnesium chloride can be formed at 10-15°C, thus minimizing the problem.

6. Commercial chlorotrimethylsilane is distilled from quinoline under nitrogen.

7. It is important that the addition be done fairly rapidly. The checkers found that slow addition (2 hr at 20°C) resulted in significant disproportionation of the ethynylmagnesium chloride.

8. It is essential to have an efficient condenser during the reflux and distillation stages because the product, trimethylsilylacetylene, is extremely volatile.

9. The receiver must be cooled to avoid serious loss of volatile product.

10. The hot distillation apparatus should be allowed to cool under a nitrogen atmosphere before being dismantled.

11. The submitters distilled the azeotrope immediately after heating the mixture to reflux. If it is allowed to stand and cool it sets solid with magnesium chloride and subsequent distillation results in appreciably lower yields.

12. The presence of residual THF in the final product is easily detected by the characteristic ^{1}H NMR signals at δ 1.85 (4 H) and 3.75 (4 H). The checkers found GLC to be more convenient than ^{1}H NMR for analysis at this point. Either a 6' x 2 mm glass column of 3% OV-101 on WHP 80/100 with 30 mL/min He or a 15 m x 0.25 mm fused silica capillary column of DB-5 (cross-linked phenylsilicone) with 1 mL/min H_2 proved sufficient to resolve butane, trimethylsilylacetylene, butyltrimethylsilane, THF, 1-chlorobutane, and bis(trimethylsilyl)acetylene.

13. The checkers used an 18-inch Vigreux column, equipped with an adjustable reflux ratio take-off. The final product can be contaminated with up to several per cent butane unless careful distillation is carried out.

14. Trimethylsilylacetylene displayed the following spectroscopic properties: IR (CCl_4) cm^{-1}: 3280 (s) and 2050 (s); ^{1}H NMR: ($CDCl_3$, 250 MHz) δ: 0.18 (s, 9 H, C$\underline{H}_3$) and 2.36 (s, 1 H, ≡C$\underline{H}$).

15. The submitters have carried out the preparation on twice the above scale with no reduction in yield.

3. Discussion

Trimethylsilylacetylene has been prepared by silylation of a variety of ethynyl metal derivatives.[2-10] The most useful methods are the silylation of ethynylmagnesium bromide[3,4,5] and chloride.[2,7,10] The use of ethynylmagnesium bromide has been reported to suffer from complicating side reactions,[2] and the results obtained in our hands were unreliable.

The present method is based on the silylation of ethynylmagnesium chloride as reported by Krüerke,[2] except that the Grignard reagent is prepared from 1-chlorobutane, rather than from the volatile (and therefore more difficult to manipulate) chloromethane. Although the preparation of ethynylmagnesium bromide is a well established Organic Syntheses procedure,[11] the use of ethynylmagnesium chloride has received little attention.[5] It does not seem to be widely appreciated that butylmagnesium chloride is much more soluble in THF than ethylmagnesium bromide and its use in Grignard ethynylations is strongly recommended. The preparation of ethynylmagnesium chloride essentially follows the Organic Syntheses procedure laid down for the bromide.[11]

Trimethylsilylacetylene is an extremely versatile (and potentially nucleophilic) two-carbon building bock. Applications in organic synthesis have been well documented in the literature.[12,13,14]

1. University Chemical Laboratory, Lensfield Road, Cambridge, CB2 1EW, U.K.
2. Krüerke, U. *J. Organomet. Chem.* **1970**, *21*, 83.
3. Kraihanzel, C. S.; Losee, M. L. *J. Organomet. Chem.* **1967**, *10*, 427.
4. Westmuze, H.; Vermeer, P. *Synthesis* **1979**, 390.
5. Brandsma, L.; Verkruijsse, H. D. "Synthesis of Acetylenes, Allenes, and Cumulenes; A Laboratory Manual"; Elsevier: New York, 1981; p. 55.
6. Bennett, G. E.; Lee, W. W., U.S. Patent 2 887 371, 1959, Monsanto Chemical Co.; *Chem. Abstr.* **1959**, *53*, 19883g.
7. Minh, L. Q.; Billiotte, J.-C.; Cadiot, P. *C.R. Hebd. Seances Acad. Sci.* **1960**, *251*, 730.
8. Buchert, H.; Zeil, W. *Z. Physik. Chem.* **1961**, *29*, 317; *Chem. Abstr.* **1962**, *56*, 4268g.

9. Shostakovskii, M. F.; Komarov, N. V.; Yarosh, O. G. *Zh. Prikl. Khim.* **1965**, *38*, 435; *Chem. Abstr.* **1965**, *62*, 13172b.

10. Cook, M. A.; Eaborn, C.; Walton, D. R. M. *J. Organomet. Chem.* **1970**, *24*, 301.

11. Skattebøl, L.; Jones, E. R. H.; Whiting, M. C. *Org. Synth., Collect. Vol. IV* **1963**, 792.

12. Cadiot, P.; Chodkiewicz, W. In "Chemistry of Acetylenes", Viehe, H. G., Ed.; Marcel Dekker: New York, 1969; Chapter 13, pp. 928-936.

13. Jäger, V.; Viehe, H. G. In Houben-Weyl, "Methoden der Organischen Chemie", Georg Thieme Verlag: Stuttgart, 1977; Vol. 5/2a, p. 390.

14. Colvin, E. "Silicon in Organic Synthesis", Butterworth: 1981, Chapter 13, pp. 165-169.

Appendix
Chemical Abstracts Nomenclature (Collective Index Number); (Registry Number)

Trimethylsilylacetylene: Silane, ethynyltrimethyl- (8,9); (1066-54-2)

Magnesium (8,9); (7439-95-4)

Iodine (8,9); (7553-56-2)

1-Chlorobutane: Butane, 1-chloro- (8,9); (109-69-3)

Acetylene (8); Ethyne (9); (74-86-2)

Chlorotrimethylsilane: Silane, chlorotrimethyl- (8,9); (75-77-4)

DIALKOXYACETYLENES: DI-tert-BUTOXYETHYNE, A VALUABLE SYNTHETIC INTERMEDIATE

(Propane, 2,2'-[1,2-ethynediylbis(oxy)]bis[2-methyl-])

A. trans-2,3-dichloro-1,4-dioxane + t-BuOH/K_2CO_3, reflux → trans-2,3-di-tert-butoxy-1,4-dioxane

B. trans-2,3-di-tert-butoxy-1,4-dioxane, 1. PCl_5/hexane, 2. liq. NH_3 → 1,2-di-tert-butoxy-1,2-dichloroethane

C. 1,2-di-tert-butoxy-1,2-dichloroethane, t-BuOK/hexane, 0° → 20° → (E)-1,2-di-tert-butoxy-1-chloroethene

D. (E)-1,2-di-tert-butoxy-1-chloroethene, $NaNH_2$/liq. NH_3, ether → t-BuO-C≡C-Ot-Bu

Submitted by Anna Bou, Miquel A. Pericàs, Antoni Riera and Fèlix Serratosa.[1]

Checked by Terry Singleton, Jae Chan Park, Martin F. Semmelhack, Sean M. Kerwin, and Clayton H. Heathcock.

1. Procedure

A. *Preparation of trans-2,3-dichloro-1,4-dioxane* (Note 1). To a 2-L, three-necked, round-bottomed flask, equipped with two inlet tubes (with sintered-glass diffusers at the end) connected to a chlorine cylinder, and a reflux condenser connected to an outlet tube immersed in a potassium hydroxide solution, are added 1200 g (13.64 mol) of anhydrous dioxane (free of peroxides!) and 8 g (0.03 mol) of iodine. A stream of chlorine is passed through the dioxane/iodine solution heated at 90°C, and the reaction is monitored by NMR spectroscopy. After 9 hr, the conversion is 50% complete (Note 2); after 33 hr, about 90% complete. At this point, the stream of chlorine is interrupted. *Reinitiation of the chlorine stream after some hours (next morning, for example) may be dangerous because it was observed in one case that the mixture inflamed spontaneously!* The reaction mixture is allowed to cool to room temperature, 500 mL of ether is added, and the solution is washed with aqueous sodium thiosulfate solution. The organic layer is separated, dried over sodium sulfate, the ether is evaporated under reduced pressure, and the residue is distilled through a 20-cm Vigreux column, to yield 1200-1300 g of trans-2,3-dichloro-1,4-dioxane, bp 89°C/16 mm (lit.[2a] bp 82.5°C/14 mm; mp 31°C) (Note 3).

2,3-Di-tert-butoxy-1,4-dioxane. To a 2-L, three-necked, round-bottomed flask, equipped with a mechanical stirrer, reflux condenser protected from moisture by a drying tube, and an inlet tube for dry nitrogen, are added 103.6 g (0.66 mol) of trans-2,3-dichloro-1,4-dioxane, 979.1 g (13.23 mol) of anhydrous tert-butyl alcohol (distilled from CaH_2), and 365.1 g (2.64 mol) of potassium carbonate (ground with a mortar and pestle and activated at 250°C for 3 hr) (Note 4). The mixture is stirred vigorously and heated under reflux for 24-30 hr, the progress of the reaction being monitored by ^1H NMR spectroscopy. Once the singlet at δ 5.95, corresponding to the methine protons of the starting material, has completely disappeared, the reaction mixture is allowed to cool to room temperature, poured into 500 mL of ether, and enough water (750-850 mL) is added to dissolve all of the inorganic salts. The organic layer is separated, and the aqueous layer is extracted with two 200-mL portions of ether. The combined ether extracts are dried over anhydrous sodium sulfate, filtered, and concentrated under reduced pressure to give 128-147 g of an oily residue (Note 5). Hexane (250 mL) is added and the solution is allowed to stand in a refrigerator. The resulting crystals are separated by suction filtration and washed thoroughly with 150-250 mL of hexane. The residue of 3-6 g of crystalline product, mp 106-107°C, which remains insoluble, is identified as trans-2-tert-butoxy-3-hydroxy-1,4-dioxane (Note 6). The filtrate (which contains approximately a 25:75 mixture of cis and trans isomers) is evaporated under reduced pressure to one-half its volume and is cooled to 0°C. Massive crystals appear which are collected by suction filtration. The crystallization process is repeated once more, to give 40-60 g of trans-2,3-di-tert-butoxy-1,4-dioxane, mp 64-65°C. The hexane solution is evaporated under reduced pressure and the oily residue distilled at 50-57°C/0.25 mm to give 54-84 g of a mixture of cis- and trans-2,3-di-tert-butoxy-1,4-dioxane (115-124 g combined; 75-81% yield) (Note 7).

B. *1,2-Di-tert-butoxy-1,2-dichloroethane.* To a 250-mL, round-bottomed flask, equipped with a pressure-equalizing dropping funnel protected from moisture by a drying-tube, and a magnetic stirring bar, are added 43.2 g (0.21 mol) of phosphorus pentachloride and 50 mL of hexane, and the flask is cooled with an ice-salt bath. While the solution is stirred, a solution of 30 g (0.13 mol) of 2,3-di-tert-butoxy-1,4-dioxane in 100 mL of hexane is added dropwise. After the addition is complete, the cooling bath is removed and the mixture is stirred at room temperature for 90-180 min until no starting material is observed in the NMR spectrum of a sample (Note 8); the reaction mixture is then filtered through a sintered-glass filter to remove excess phosphorus pentachloride. The resulting hexane solution (~ 200 mL), which contains 1,2-di-tert-butoxy-1,2-dichloroethane, 2-chloroethyl dichlorophosphate, 1,2-dichloroethane, phosphorus oxychloride and traces of phosphorus pentachloride, is transferred into a 1-L, three-necked, round-bottomed flask, equipped with a magnetic stirrer, a dry-ice condenser protected from moisture by a potassium hydroxide tube, and a short inlet tube connected to an ammonia cylinder. The reaction flask is cooled with a dry ice/acetone bath and, while the solution is stirred (Note 9), a fast stream of gaseous ammonia is introduced. A vigorous reaction takes place and a copious white precipitate forms. The stream of gaseous ammonia is continued for 15-60 min (Note 10). The cooling bath and condenser are removed, the reaction flask is connected to an ordinary aspirator line through a potassium hydroxide drying trap, and the ammonia is evaporated under aspirator vacuum with efficient stirring. The ammonia-free solution is filtered through a sintered-glass filter and the precipitate is washed with hexane. The resulting hexane solution (400-500 mL), which contains 25-26 g (0.12 mol) of pure 1,2-di-tert-butoxy-1,2-dichloroethane, is suitable for the next operation (yields, calculated from an aliquot, are 95-97%) (Note 11).

C. *(E)-1,2-Di-tert-butoxy-1-chloroethene*. The solution prepared above is placed in a 1-L, round-bottomed flask, equipped with a magnetic stirring bar and a Liebig condenser (protected from moisture by a drying tube) and cooled with an ice bath; 28.9 g (0.258 mol) of solid potassium tert-butoxide is added in small portions through the condenser over a 30-min period. After addition, the cooling bath is removed and stirring is continued for 90-120 min until no more starting material is detected in the ^{1}H NMR spectrum of a sample; enough ice water is then added to just dissolve all the inorganic salts. The organic layer is separated and the aqueous layer is extracted with hexane (2 x 100 mL). The combined hexane extracts are dried over anhydrous sodium sulfate, filtered, and concentrated at aspirator vacuum. The residue is distilled at 40°C/0.2 mm, with the collection flask at -78°C, to give a center cut of (E)-1,2-di-tert-butoxy-1-chloroethene (15.4-17.0 g, 58-63% yield from 2,3-di-tert-butoxy-1,4-dioxane) (Notes 12, 13).

D. *Di-tert-Butoxyethyne*. In a 2-L, three-necked, round-bottomed flask, equipped with a magnetic stirring bar (Note 14), dry-ice condenser protected from moisture by a potassium hydroxide tube, and a pressure-equalizing dropping funnel, 0.5 mol of sodium amide is prepared in 500 mL of liquid ammonia (Note 15), and 20 g (0.0968 mol) of (E)-1,2-di-tert-butoxy-1-chloroethene dissolved in 150 mL of anhydrous ether is added in a 5-min period with efficient stirring. After the addition, stirring is continued for 80 min. The reaction mixture is diluted with 200 mL of cold pentane (-20°C to -30°C), and 400 mL of cold water is added very cautiously. The organic layer is washed with 50 mL of a 0.1 M buffered phosphate solution (NaH_2PO_4/Na_2HPO_4), dried over anhydrous sodium sulfate (Note 16), filtered and concentrated at aspirator vacuum without heating to give 10-12 g (78-86% yield) of di-tert-butoxyethyne as a pale yellow oil. The product is sufficiently pure for further reactions (Note 17), but it may be distilled at 30°C/0.05 mm; freezing point 8.5°C.

2. Notes

1. This procedure was reported J. J. Kucera and D. C. Carpenter.[2]

2. These data were obtained by the checkers. The submitters report conversion of 76-80% after only 9 hr. It seems likely that the rate of the reaction may be sensitive to the dimensions and mechanical features of the chlorine introduction system, and/or an induction period. It is easy and important to monitor the process.

3. trans-2,3-Dichloro-1,4-dioxane has the following spectra; IR (CCl_4) cm^{-1}: 2990, 2940, 2885, 1455, 1385, 1375, 1337, 1160, 1115, 1032, 900, 875, 670; ^{1}H NMR (CCl_4)[2b] δ: 3.40-4.57 (AA'BB', 4 H, C$\underline{H}_2$), 5.95 (s, 2 H, ClC$\underline{H}$O).

4. The yields of this reaction are very sensitive to the presence of traces of moisture, and to the ratio of reagents. If one works without a nitrogen atmosphere and with tert-butyl alcohol which has not been previously dried over calcium hydride, with a 1:10 ratio of dichloro derivative/alcohol, the yields drop to 65%.

5. *Caution should be exercised in evaporation of the ether, as the di-tert-butoxy compounds are appreciably volatile at reduced pressure. If a rotary evaporator is used for the concentration, the water bath should be kept at or below room temperature, and the residue should not be pumped after it is clear that the bulk of the ether has been evaporated.* The submitters report ca. 130 g of oily residue at this stage. Treatment of the oily residue with 105 mL of ether leads to partial crystallization. During washing of the crystals in a Büchner filter with more ether, an almost complete solubilization takes place, but eventually 0.1-0.5 g of cis-2,3,7,10-tetraoxabicyclo[4.4.0]decane remains as an insoluble residue. This compound was prepared for the first time in 1931;[3] the cis configuration was only established in 1966.[4] It has the following properties: mp 136°C; IR (CCl_4) cm^{-1}: 2980, 2955, 2930, 2910, 2875, 1460, 1350, 1285, 1260, 1250, 1150, 1140, 1095, 1080, 1025, 910, 870, 780; ^{1}H NMR (CCl_4) δ: 3.30-4.20 (AA'BB', 8 H, C$\underline{H}_2$), 4.60 (s, 2 H, OC$\underline{H}$O).

6. The crystalline trans isomer epimerizes in chloroform solution to give a nearly 70:30 mixture of cis and trans isomers. The trans isomer, mp 106-107°C, shows the following spectroscopic properties: IR (KBr) cm^{-1}: 3450, 2980, 2935, 2890, 1445, 1390, 1370, 1335, 1280, 1260, 1200, 1135, 1105,

1060, 1045, 1035, 1020, 910, 855, 780; ^{1}H NMR (CCl$_4$) δ: 1.27 (s, 9 H, C$\underline{H}_3$), 3.33-4.13 (ABCD + OH, 4 H + 1 H) 4.57 (br, 2 H, OC$\underline{H}$O). Anal. Calcd for C$_8$H$_{16}$O$_4$: C, 54.53; H. 9.15. Found: C, 54.53; H 9.29.

7. trans-2,3-Di-tert-butoxy-1,4-dioxane has the following spectra: IR (CCl$_4$) cm^{-1}: 2975, 2930, 1390, 1367, 1190, 1145, 1100, 1060, 1040, 857; ^{1}H NMR (CCl$_4$) δ: 1.19 (s, 18 H, C$\underline{H}_3$), 3.05-4.20 (m, AA'BB', 4 H, C$\underline{H}_2$), 4.30 (s, 2 H, OC$\underline{H}$O). cis- + trans-2,3-Di-tert-butoxy-1,4-dioxane have the following additional signals: IR (CCl$_4$) cm^{-1}: 1170, 1130, 1120, 1080, 1020, 1000, 960, 879; ^{1}H NMR (CCl$_4$) δ: 4.43 (s, 2 H, OC$\underline{H}$O, cis).

8. The peaks corresponding to 2,3-di-tert-butoxy-1,4-dioxane overlap with those of the methylene protons of 2-chloroethyl dichlorophosphate, a by-product from the reaction, but the absence of the acetal protons of the starting material is clear from the symmetry of the multiplet.

9. The solution is thick at -78°C. Dilution with additional hexane may be necessary.

10. The end of the reaction can be easily detected because, when all of the 2-chloroethyl dichlorophosphate has been destroyed, the reaction mixture *does not crackle* any more when condensed ammonia drops on the stirred mixture, or, much more easily, when the reaction mixture becomes basic to pH paper.

11. The crude reaction mixture is, in fact, an approximately 30:70 mixture of dl- and meso-1,2-di-tert-butoxy-1,2-dichloroethane. Although the ^{1}H NMR spectrum at 60 MHz (CCl$_4$) shows only one singlet at 5.6 ppm, the 200 MHz spectrum (CDCl$_3$) shows two sharp singlets separated by 1.8 Hz. The pure meso compound could be isolated by crystallization and purified by sublimation at 40°C/0.05 mm; mp 77-78° (dec). The spectra are as follows: IR (CCl$_4$) cm^{-1}: 2975, 2925, 1470, 1458, 1390, 1368, 1310, 1250, 1180, 1130, 1025, 850, 650; ^{1}H NMR at 200 MHz (CDCl$_3$) δ: 1.36 (s, 9 H, C$\underline{H}_3$), 5.73 (s, 1 H, OC$\underline{H}$Cl).

12. In later fractions, small amounts (0.2-0.5%) of (Z)-1,2-di-tert-butoxy-1-chloroethene have been detected: ^{1}H NMR (CCl$_4$) δ: 1.25 (s, 9 H C$\underline{H}_3$), 1.33 (s, 9 H, C$\underline{H}_3$), 6.03 (s, 1 H, =C$\underline{H}$).

13. (E)-1,2-Di-tert-butoxy-1-chloroethene has the following properties: n_D^{25} 1.4410-1.4415; UV (cyclohexane): 217.7 nm (log ε = 3.7); IR (CCl$_4$) cm^{-1}: 2972, 1670, 1470, 1390, 1366, 1290, 1260, 1240, 1180, 1140, 1070, 1025, 935; ^{1}H NMR (CCl$_4$) δ: 1.26 (s, 9 H, C$\underline{H}_3$), 1.33 (s, 9 H, C$\underline{H}_3$), 5.91 (s, 1 H, =C$\underline{H}$).

14. The stirring bar must be glass-covered, since sodium in ammonia solution attacks Teflon.

15. The method used for the preparation of sodium amide is a modification of the procedure described by Nieuwland, Vaughn, and Vogt.[5] In a 3-L, three-necked, round-bottomed flask, equipped with a magnetic stirring bar (Note 14), a dry-ice condenser protected from moisture by a potassium hydroxide tube, and an inlet tube connected to the ammonia cylinder, is condensed 500 mL of liquid ammonia. A slow stream of dry oxygen is initiated through the inlet tube and 11.5 g (0.5 mol) of sodium in small pieces is slowly introduced. The addition of sodium requires 4-5 hr, since the blue color must be discharged before each new addition of sodium. In this way, a completely white suspension of sodium amide is obtained, which allows the formation of crude di-tert-butoxyethyne, free from any iron impurities.

16. It is best to minimize exposure of di-tert-butoxyethyne to light.

17. Eventually, if a more concentrated solution of (E)-1,2-di-tert-butoxy-1-chloroethene in ether is used, the formation of 1,2,3-tri-tert-butoxy-3-cyano-1-propene [^{1}H NMR (CCl$_4$) δ: 6.05 (s, 1 H), 4.98 (s, 1 H), 1.28 (br, 27 H)] and 1,2,3-tri-tert-butoxy-1-cyano-1-propene [^{1}H NMR (CCl$_4$) δ: 4.03 (s, 2 H), 1.28 (b, 27 H)] is observed. The by-products may be eliminated by

column chromatography on neutral alumina (40 g, 100-125 mesh, activity 1), using a column refrigerated at 0°C and protected from the light, and eluting with pentane under nitrogen pressure. From the first 750 mL of eluant, 9-12 g of pure di-tert-butoxyethyne is obtained.

18. Di-tert-butoxyethyne has the following properties: n_D^{25} 1.4365; IR (CCl_4) cm^{-1}: 2972, 2922, 1470, 1450, 1390, 1367, 1301, 1263, 1245, 1150, 825; ^1H NMR (CCl_4) δ: 1.31 (s, C$\underline{H}_3$).

3. Discussion

The present procedure for the preparation of di-tert-butoxyethyne is an improvement of a method previously reported by the submitters,[6] who have also reported the preparation from glyoxal via 1,2-dichloro-1,2-dimethoxyethane.[6]

Although acetylenic diethers are thermodynamically stable compounds, they show a high kinetic instability that induces polymerization even at low temperatures.[7]

Different acetylenic diethers have been prepared, either from glyoxal (method B) or from dioxane (method A); their stability correlates well, in a qualitative way, with Charton's ν steric parameter,[8] based on effective Van der Waals radii for the corresponding alkoxy groups:

ACETYLENIC DIETHERS

RO-C≡C-OR

-OR	Method	ν	$t_{1/2}$ Order of magnitude (by NMR)
-OCH$_3$	B[9]	0.38	seconds (0°C, soln.)
-OC$_2$H$_5$	B[7]	0.48	seconds (0°C, soln.)
-OCH$_2$C(CH$_3$)$_3$	A[10]	0.70	minutes (r.t., soln.)
-OCH(CH$_3$)$_2$	A,B[6]	0.75	minutes (r.t., soln.)
-OC(CH$_3$)$_3$	A,B[6]	1.22	days (r.t., neat)

As shown above, di-tert-butoxyethyne is the only acetylenic diether prepared so far whose stability allows its use as a synthetic intermediate. It has been used in the synthesis of all the members of the series of monocyclic oxocarbons (deltic, squaric, croconic, and rhodizonic acids),[11] as well as in the synthesis of semisquaric acid, the parent compound of the natural mycotoxins, *moniliformins*.[12]

Di-tert-butoxyethyne, like other acetylenic ethers having hydrogen atoms at the position, is thermally unstable, and when a benzene solution is heated under reflux, elimination occurs to give isobutene and tert-butoxyketene. tert-Butoxyketene then reacts with the parent acetylene to afford 2,3,4-tri-tert-butoxycyclobutenone, which is the precursor of squaric and semisquaric acid.[11,12] This thermal instability prevents the use of di-tert-butoxyethyne in those reactions that proceed at temperatures higher than 40-50°C, such as some Diels-Alder condensations.

On the other hand, di-tert-butoxyethyne is prone to undergo a variety of reactions with metal transition complexes [$PdCl_2 \cdot CH_3CN$, $Co_2(CO)_8$, $CpCo(CO)_2$, $Ni(CO)_4$], to afford new, transient, intermediate complexes which are the actual precursors of oxocarbons.[13]

1. Departament de Química Orgànica, Facultat de Química, Universitat de Barcelona, Diagonal 647, E-08028 Barcelona, Spain.
2. (a) Kucera, J. J.; Carpenter, D. C. *J. Am. Chem. Soc.* **1935**, *57*, 2346; (b) Jung, D. *Chem. Ber.* **1966**, *99*, 566.
3. Böeseken, J.; Tellegen, F.; Cohen Henriquez, P. *Recl. Trav. Chim. Pays-Bas* **1931**, *50*, 909.
4. Altona, C.; Havinga, E. *Tetrahedron* **1966**, *22*, 2275; Fraser, R. R.; Reyes-Zamora, C. *Can. J. Chem.* **1967**, *45*, 929.
5. Vaughn, T. H.; Vogt, R. R.; Nieuwland, J. A. *J. Am. Chem. Soc.* **1934**, *56*, 2120.
6. Bou, A.; Pericàs, M. A.; Serratosa, F. *Tetrahedron* **1981**, *37*, 1441.
7. Pericàs, M. A., Doctoral Thesis, University of Barcelona, 1979.
8. Charton, M. *Progr. Phys. Org. Chem.* **1973**, *10*, 81; Charton, M. *J. Org. Chem.* **1978**, *43*, 3995.
9. Messeguer, A.; Serratosa, F.; Rivera, J. *Tetrahedron Lett.* **1973**, 2895.
10. Pericàs, M. A.; Riera, A.; Serratosa, F. *Tetrahedron* **1982**, *38*, 1505.
11. Pericàs, M. A.; Serratosa, F. *Tetrahedron Lett.* **1977**, 4437.
12. Bou, A.; Pericàs, M. A.; Serratosa, F. *Tetrahedron Lett.* **1982**, *23*, 361.
13. Serratosa, F. *Acc. Chem. Res.* **1983**, *16*, 170.

Appendix

Chemical Abstracts Nomenclature (Collective Index Number);

(Registry Number)

Di-tert-butoxyethyne: Propane, 2,2'-[1,2-ethynediylbis(oxy)]bis[2-methyl- (10); (66478-63-5)

trans-2,3-Dichloro-1,4-dioxane: p-Dioxane, 2,3-dichloro-, trans- (8); 1,4-Dioxane, 2,3-dichloro-, trans- (9); (3883-43-0)

Dioxane: p-Dioxane (8); 1,4-Dioxane (9); (123-91-1)

Chlorine (8,9); (7782-50-5)

tert-Butyl alcohol (8); 2-Propanol, 2-methyl- (9); (75-65-0)

trans-2,3-Di-tert-butoxy-1,4-dioxane: 1,4-Dioxane, 2,3-bis(1,1-dimethylethoxy)-, trans- (10); (68470-79-1)

cis-2,3-Di-tert-butoxy-1,4-dioxane: 1,4-Dioxane, 2,3-bis(1,1-dimethylethoxy)-, cis- (10); (68470-78-0)

dl-1,2-Di-tert-butoxy-1,2-dichloroethane: Propane, 2,2'-[(1,2-dichloro-1,2-ethanediyl)]bis(oxy)bis[2-methyl-, (R*,R*)-(±)- (10); (68470-80-4)

meso-1,2-Di-tert-butoxy-1,2-dichloroethane: Propane, 2,2'-[(1,2-dichloro-1,2-ethanediyl)bis(oxy)]bis[2-methyl-, (R*,S*)- (10); (68470-81-5)

Phosphorus pentachloride: Phosphorus chloride (8); Phosphorane, pentachloro- (9); (10026-13-8)

2-Chloroethyl dichlorophosphate: Phosphorodichloridic acid, 2-chloroethyl ester (8,9); (1455-05-6)

1,2-Dichloroethane: Ethane, 1,2-dichloro- (8,9); (107-06-2)

Phosphorus oxychloride: Phosphoryl chloride (8,9); (10025-87-3)

Ammonia (8,9); (7664-41-7)

(E)-1,2-Di-tert-butoxy-1-chloroethene: Propane, 2,2'-[(1-chloro-1,2-ethenediyl)bis(oxy)]bis[2-methyl-, (E)- (10); (70525-93-8)

Potassium tert-butoxide: tert-Butyl alcohol, potassium salt (8); 2-Propanol, 2-methyl-, potassium salt (9); (865-47-4)

Sodium amide (8,9); (7782-92-5)

Sodium (8,9); (7440-23-5)

cis-2,5,7-10-Tetraoxabicyclo[4.4.0]decane: p-Dioxino[2,3,-b]-p-dioxin, hexahydro- (8); [1,4]-Dioxino[2,3-b]-1,4-dioxin, hexahydro (9); (4362-05-4)

METHYLENATION OF CARBONYL COMPOUNDS: (+)-3-METHYLENE-cis-p-MENTHANE

(Cyclohexane, 4-methyl-2-methylene-1-(1-methylethyl)-, R,R-)

A. [structure: (+)-isomenthol] $\xrightarrow[\text{25°C}]{\text{K}_2\text{Cr}_2\text{O}_7, \text{H}_2\text{SO}_4 \atop \text{ether - H}_2\text{O}}$ [structure: (+)-isomenthone]

B. [structure: (+)-isomenthone] $\xrightarrow[\text{room temp.}]{\text{Zn - CH}_2\text{Br}_2 - \text{TiCl}_4 \atop \text{THF - CH}_2\text{Cl}_2}$ [structure: methylenated product]

Submitted by Luciano Lombardo.[1]
Checked by Mamoru Uchiyama and Ryoji Noyori.

1. Procedure

A. *(+)-Isomenthone.* Into a 1-L, three-necked, round-bottomed flask equipped with a mechanical stirrer, condenser, thermometer and dropping funnel, are placed 54.6 g (0.35 mol) of (+)-isomenthol (Note 1) and 350 mL of ether. A solution of chromic acid, prepared by mixing 56.7 g (0.23 mol) of potassium dichromate and 31.0 mL (0.58 mol) of 98% sulfuric acid and diluting to 200 mL with water, is added dropwise to maintain the reaction temperature at 25°C. The mixture is stirred for a further 2 hr. The ether layer is separated and the aqueous phase is extracted twice with 100-mL portions of ether. The combined ether extracts are washed with saturated sodium bicarbonate solution, dried over magnesium sulfate, and evaporated under reduced pressure to leave an oil. Distillation through a short Vigreux column gives the main fraction of (+)-isomenthone as a clear colorless liquid (40.5 g), bp 64-64.5°C/5 mm; $[\alpha]_D^{16}$ +114° [$CHCl_3$, c 5.09].

B. *(+)-3-Methylene-cis-p-menthane*. Into a 1-L, round-bottomed flask fitted with a magnetic stirrer and a pressure equalizing dropping funnel connected to a nitrogen line are placed 28.75 g (0.44 mol) of activated zinc powder (Note 2), 250 mL of dry tetrahydrofuran (Note 3), and 10.1 mL (0.144 mol) of dibromomethane (Note 4). The mixture is stirred and cooled with a dry ice/acetone cooling bath at -40°C. To the stirred mixture is added dropwise 11.5 mL (0.103 mol) of titanium tetrachloride (Note 4) over 15 min. The cooling bath is removed and the mixture is stirred (Note 5) at 5°C (cold room) for 3 days under a nitrogen atmosphere. The dark grey slurry (Note 6) is cooled with an ice/water bath and 50 mL of dry dichloromethane (Note 7) is added. To the stirred mixture is added 15.4 g (0.1 mol) of (+)-isomenthone in 50 mL of dry dichloromethane over a period of 10 min. The cooling bath is removed and the mixture is stirred at room temperature (20°C) for 1.5 hr. The mixture is diluted with 300 mL of pentane and a slurry of 150 g of sodium bicarbonate in 80 mL of water is added cautiously (Note 8) over 1 hr. The clear organic solution is poured off into a 1.5-L Erlenmeyer flask and the residue is washed three times with 50-mL portions of pentane. The combined organic solutions are dried over a mixture of 100 g of sodium sulfate and 20 g of sodium bicarbonate (Note 9), filtered through a sintered glass funnel (No. 2), and the solid desiccant is thoroughly washed with pentane. The solvent is removed at atmospheric pressure by flash distillation through a column (40 cm x 2.5 cm) packed with glass helices. The liquid residue is distilled (Note 10) to give the methylenated product as a clear, colorless liquid, bp 105-107°C/90 mm, 13.6 g, 89% yield (Note 11), n_D^{24} 1.45321, $[\alpha]_D^{23}$ +7.7 to +8.6° [$CHCl_3$, c 4.0].

2. Notes

1. (+)-Isomenthol was purchased from the Aldrich Chemical Company, Inc. Oxidation was carried out according to the procedure of Brown.[2]

2. Zinc powder from Hopkin and Williams Chemical Company or Nakarai Chemicals (GR grade) is activated according to Fieser and Fieser.[3]

3. Tetrahydrofuran is dried by distillation from sodium/benzophenone.

4. Dibromomethane from EGA-CHEMIE and titanium tetrachloride from E. Merck are used as supplied. The checkers used the products of Nakarai Chemicals. All residues of titanium tetrachloride are destroyed with acetone from a wash bottle.

5. As the reaction progresses the mixture thickens and it is necessary to begin with a reasonably fast rate of stirring. However, too fast a stirring rate causes the mixture to splash up to the neck of the round-bottomed flask as the mixture thickens.

6. The reagent must be kept cold at all times because at room temperature the active reagent slowly decomposes and the mixture darkens considerably. Once prepared the reagent can be stored at -20°C (freezer) in a well-sealed flask without a significant loss of activity. A sample stored in this way for 1 year showed only a slight, ~ 5-10%, loss of activity. The molar activity of the active reagent is equivalent to the titanium tetrachloride ($TiCl_4$) molarity (determined by reaction with excess ketone followed by GLC analysis); however, an increase in the proportion of $TiCl_4$ makes no difference to the molar activity.

7. Analytical Reagent ANALAR dichloromethane was dried by storing over 4Å sieves. The checkers purchased the EP grade solvent from Wako Pure Chemical Industries.

8. It is necessary to add the slurry dropwise at the beginning, allowing the effervescence to subside after each drop. After the initial vigorous effervescence, larger portions can be added. During this part of the addition, the stirrer becomes ineffective and gentle shaking by hand is continued until effervescence ceases.

9. The organic solution is shaken with the drying agent for 10-15 min to remove the last traces of titanium salts.

10. The pressure is allowed to drop to ~ 40 mm for a few minutes when the oil bath temperature has reached 50°C. This procedure removes any residual tetrahydrofuran which can be responsible for a small contaminated forerun. A considerable amount of product is collected at the end of the distillation as the temperature drops.

11. This is the first preparation of this compound (see reference 9 of discussion); the data obtained are as follows: Anal. Calcd for $C_{11}H_{20}$: C, 86.83, H, 13.48. Found: C, 86.76, H, 13.24. ^{1}H NMR (CDCl$_3$, 200 MHz) δ: 0.79 (d, 3 H, J = 7, -C$\underline{H}_3$); 0.91 (d, 6 H, J = 7, -C$\underline{H}_3$); 1.01-2.14 (m, 9 H, -C$\underline{H}_2$-, -C$\underline{H}$); 4.54, 4.60, (two s, 2 H, =C$\underline{H}_2$); ^{13}C NMR (CDCl$_3$): 17.4, 18.3, 19.2, 22.6, 25.8, 26.5, 31.8, 37.3, 47.4, 104.6, 148.4.

3. Discussion.

This new Zn/CH$_2$Br$_2$/TiCl$_4$ procedure[4] provides a mild, non-basic method for the methylenation of ketones (a competitive pinacol dimerization occurs with aldehydes but good yields of the olefin can still be achieved) and offers an important alternative to the standard Wittig[5] reaction. These characteristics are derived from two important observations:

(i) Ketones are not enolized by the reagent and one important consequence is that adjacent enolizable chiral centers are not epimerized.

(ii) The reagent is compatible with a wide variety of functional groups, for example (Table I), THP ethers, tert-butyldimethylsilyl ethers, acetals, esters, carboxylic acids, alcohols, lactones. Such selectivity makes it a valuable procedure in organic synthesis[6] and appreciably augments Wittig methodology.

The corresponding Wittig reagent, $CH_2=PPh_3$, reacts smoothly with both aldehydes and ketones to give methylenated products in high yield but with one subtle limitation. The problem cannot be detected with aldehydes because they react rapidly even at temperatures as low as -78°C, but ketones react more slowly, and an adjacent enolizable chiral center can be epimerized as a result of competitive reversible enolization. This limitation of the Wittig procedure has been recognized for some time[7] and new methylenation methods[7,8] that avoid enolization have been developed. However, the application of these methods is restricted either by low yields or by incompatibility with other functionality in the molecule. In the closest analogy to the present preparation of (+)-3-methylene-cis-p-menthane, one of these methods has been used for the methylenation of ℓ-menthone[9] but a yield of only 40% was obtained.

The $Zn/CH_2Br_2/TiCl_4$ procedure as originally reported[10] has not been widely used. The active reagent was generated and trapped by the carbonyl compound in situ at room temperature. Long reaction times were required and as a consequence the substrate was exposed for long periods to $TiCl_4$, severely limiting the usefulness of the procedure. The discovery[4] that the active reagent was stable at low temperature and could be preformed extends the utility and scope of the reaction enormously. This reagent reacts rapidly with ketones at room temperature with considerably improved yields and selectivities.

The non-basic nature of the reagent makes it useful in other applications. It has, for example, also proved useful for the methylenation of the gibberellin norketone[4] (Table I) without the need for protection of the readily epimerized 3β-OH. The use of CD_2Br_2[4] allows the introduction of $=CD_2$ without scrambling of the label.

The $Zn/CH_2Br_2/TiCl_4$ reagent is superior to other non-basic methylenation reagents. Furthermore the long shelf-life at low temperatures, together with the ease of workup, render it an appealing alternative to the Wittig method for the methylenation of ketones in general.

1. Research School of Chemistry, The Australian National University, G.P.O. Box 4, Canberra, A.C.T. 2601, Australia. Financial support from the Queen Elizabeth II Fellowship Committee and technical support by Mr. P. Lyndon for preparation A is gratefully acknowledged.
2. Brown, H. C.; Garg, C. P. *J. Am. Chem. Soc.* **1961**, *83*, 2952.
3. Fieser, L. F.; Fieser, M. "Reagents for Organic Synthesis"; Wiley: New York, 1967; Vol. I, p. 1276.
4. Lombardo, L. *Tetrahedron Lett.* **1982**, *23*, 4293.
5. Maercker, A. *Org. React.* **1965**, *14*, 270; Johnson, A. W. "Ylid Chemistry", Academic Press: New York, 1966.
6. (a) Gibberellin synthesis: Lombardo, L.; Mander, L. N. *J. Org. Chem.* **1983**, *48*, 2298; (b) Prostaglandin synthesis: Shibasaki, M.; Torisawa, Y.; Ikegami, S. *Tetrahedron Lett.* **1983**, *24*, 3493; Ogawa, Y.; Shibasaki, M. *Tetrahedron Lett.* **1984**, *25*, 1067.

7. Sowerby, R. L.; Coates, R. M. *J. Am. Chem. Soc.* **1972**, *94*, 4758 and references therein.
8. Hasselmann, D. *Chem. Ber.* **1974**, *107*, 3486; Meyers, A. I.; Ford, M. E. *Tetrahedron Lett.* **1975**, 2861.
9. Shono, T.; Matsumura, Y.; Kashimura, S.; Kyutoku, H. *Tetrahedron Lett.* **1978**, 2807. The stereochemistry of the starting menthone is not defined in the article, but in a personal communication Professor Shono has revealed that it was ℓ-menthone and the accompanying spectral data of the methylenated product differed from that of the (+)-3-methylene-cis-p-menthane obtained in this procedure. The same trans-p-menthane product as that of T. Shono et al. was obtained, using a different method, by Seitz, D. E.; Zapata, A. *Tetrahedron Lett.* **1980**, *21*, 3451. The compound obtained in both of these articles has been incorrectly cited in *Chem. Abstr.* **1979**, *90*, 71804j; *Chem. Abstr.* **1981**, *94*, 174221d as possessing the cis-p-menthane skeleton. Similarly, the abstract, *Chem. Abstr.* **1979**, *91*, 20016s has incorrectly cited the cis-p-menthane skeleton for the product resulting from ℓ-menthone.
10. Takai, K.; Hotta, Y.; Oshima, K.; Nozaki, H. *Tetrahedron Lett.* **1978**, 2417; Takai, K.; Hotta, Y.; Oshima, K.; Nozaki, H. *Bull. Chem. Soc. Jpn.* **1980**, *53*, 1698.

Table I. Methylenation of Ketones with $Zn/CH_2Br_2/TiCl_4$

Substrate		Isolated Yield, %	Ref.
(gibberellin-type structure with HO, CH₃, CO₂R¹, R²)	$R^1 = R^2 = H$	80	2
	$R^1 = CH_3, R^2 = H$	93	
(decalin with HO, dioxolane, CO₂CH₃)		90	2
(cyclopentanone with THPO, OTHP, CO₂CH₃ side chain)		80	4b
(cyclopentanone with THPO, OSi(CH₃)₃-type, CO₂CH₃ side chain)		81	4b

Appendix

Chemical Abstracts Nomenclature (Collective Index Number);

(Registry Number)

(+)-Isomenthone: Cyclohexanone, 5-methyl-2-(1-methylethyl)-, (2R-cis)- (9); (1196-31-2)

(+)-Isomenthol: Cyclohexanol, 5-methyl-2-(1-methylethyl)-, [1S-(1α,2β,5β)]- (9); (23283-97-8)

Potassium dichromate: Dichromic acid, dipotassium salt (8); Chromic acid, dipotassium salt (9); (7778-50-9)

Zinc (8,9); (7440-66-6)

Dibromomethane: Methane, dibromo- (8,9); (74-95-3)

Titanium tetrachloride: Titanium chloride (8,9); (7550-45-0)

Dichloromethane: Methane, dichloro- (8,9); (75-09-2)

A GENERAL SYNTHETIC METHOD FOR THE PREPARATION OF CONJUGATED DIENES FROM OLEFINS USING BROMOMETHANESULFONYL BROMIDE: 1,2-DIMETHYLENECYCLOHEXANE

(Cyclohexane, 1,2-bis(methylene)-)

A. sym-trithiane + 9 Br_2 $\xrightarrow{H_2O}$ 3 $BrCH_2SO_2Br$

B. cyclohexene $\xrightarrow[h\nu]{BrCH_2SO_2Br}$ (trans-1-bromo-2-(bromomethylsulfonyl)cyclohexane) $\xrightarrow{KOt\text{-}Bu}$ 1,2-dimethylenecyclohexane

Submitted by Eric Block and Mohammad Aslam.[1]
Checked by Jeffrey C. Weber and Leo A. Paquette.

1. Procedure

A. Bromomethanesulfonyl bromide. A 3-L, three-necked, round-bottomed flask equipped with a mechanical stirrer, pressure-equalized dropping funnel, and a thermometer is charged with 100 g (0.73 mol) of sym-trithiane (Note 1) suspended in 600 mL of water. Bromine (1136 g, 7.1 mol) is added with stirring while keeping the flask temperature around 40°C (Note 2). After the addition of half of the bromine, 600 mL of water is added and bromine addition is continued. After all of the bromine has been added, the reaction mixture is stirred for 0.25 hr. The mixture is transferred to a 4-L separatory funnel, the lower organic layer is separated, and the aqueous layer is extracted with two 200-mL portions of methylene chloride (Note 3). The organic extracts are combined, washed with one 100-mL portion of cold 5% sodium bisulfite solution and with 100 mL of water, dried over anhydrous magnesium sulfate, and concentrated at room temperature with a rotary evaporator to a light yellow oil. Distillation using a short Vigreux column affords 218-249 g (42-48%) of bromomethanesulfonyl bromide as a light yellow oil, bp 68-69°C (0.55 mm) (Notes 4, 5).

B. *1-Bromo-1-methyl-2-(bromomethylsulfonyl)cyclohexane*. Four Pyrex test tubes (2.5 x 20 cm) are charged with 1-methylcyclohexene (5.0 g per test tube; total weight 20.0 g, 0.21 mol) (Note 6). Methylene chloride (12 mL) is added to each test tube which is cooled in ice. An ice-cold solution of bromomethanesulfonyl bromide (13.6 g of bromomethanesulfonyl bromide per test tube; total weight 54.4 g, 0.23 mol) in methylene chloride (12 mL) is added to each test tube with mixing at 0°C (Note 7). The test tubes are attached with the help of several rubber bands to a Pyrex immersion well equipped with a 450-W mercury lamp (Note 8). The immersion well is cooled by circulation of ice water (Note 9) and immersed in a cooling bath maintained at -15°C. The reaction mixture is irradiated for 2 hr. Solid potassium carbonate (1.5 g) is added to each test tube and the contents of the test tubes are filtered through a small column with a glass wool plug into a 250-mL round-bottomed flask. Methylene chloride is removed, first on a rotary evaporator and then with a vacuum pump (1 mm), to give an oil which gradually solidifies (68.3 g, 98%). Recrystallization from 95% ethanol (100 mL) gives white crystals (54.3 g, 78%) (Note 10), mp 59-61°C (Note 11).

C. *1,2-Dimethylenecyclohexane*. An oven-dried, 1-L, three-necked, round-bottomed flask equipped with a mechanical stirrer, pressure-equalized dropping funnel, and a stopper is charged with potassium tert-butoxide (59.5 g, 0.53 mol) (Note 12) dissolved in tert-butyl alcohol-tetrahydrofuran (9:1, 400 mL total) (Note 13) and cooled in ice. A solution of 1-bromo-1-methyl-2-(bromomethylsulfonyl)cyclohexane (54.0 g, 0.16 mol) in tert-butyl alcohol-tetrahydrofuran (9:1, 100 mL) (Note 14) is added dropwise over a 1-hr period. After the addition is complete the reaction mixture is stirred at room temperature for 0.5 hr and then poured into a 2-L separatory funnel containing 500 mL of water. This solution is extracted with two 150-mL portions of pentane. The combined pentane extracts are washed eight times with water (500 mL) (Note 15), dried over anhydrous magnesium sulfate and filtered. The pentane is removed by distillation at atmospheric pressure using an efficient Vigreux column and the residue is distilled under reduced pressure to give 11.4 g (65%) of 1,2-dimethylenecyclohexane as a colorless liquid, bp 69-70°C (90 mm) (lit[2] 60-61°C, 90 mm) (Notes 16 and 17).

2. Notes

1. Sym-Trithiane is made as described by Bost and Constable[3] and is used without purification. It can be purchased from Aldrich Chemical Company, Inc.

2. This is an exothermic reaction; no outside heating is required. If the temperature goes above 40°C, the flask is cooled by ice/water.

3. In the first of these extractions, the upper phase is the organic one. In the second extraction, the organic layer is at the bottom.

4. The product has refractive index n_D^{20} 1.5706 and spectral properties as follows: IR (neat) cm^{-1}: 3040 (vs), 2960 (vs), 1362 (vs), 1205 (s), 1160 (vs), 1105 (m), 830 (s), 680 (s); ^1H NMR ($CDCl_3$, 60 MHz) δ: 5.05 (s).

5. The product can be synthesized on a much smaller scale with no loss in yield simply by reducing the quantities as desired.

6. 1-Methylcyclohexene was obtained from the Aldrich Chemical Company, Inc. and was distilled before use.

7. On occasion bromomethanesulfonyl bromide undergoes spontaneous, exothermic addition to olefins. While this problem was not encountered with 1-methylcyclohexene, it is desirable to mix the reagents at low temperature to avoid a possible vigorous spontaneous reaction and to maximize the yield of adduct.

8. The apparatus should be shielded to avoid exposure to ultraviolet light. The immersion well, lamp, and the requisite transformer are available from Hanovia Lamp Division, Canrad-Hanovia Inc., 100 Chestnut Street, Newark, NJ 07105. The test tubes are positioned so as to be as close as possible to the lamp.

9. A "Little Giant" submersible pump (available from Little Giant Pump Co., Oklahoma City, OK) is used to circulate ice water through the immersion well.

10. The first crop (47.0 g) is followed by two other crops (5.2 and 2.1 g) obtained by concentrating and cooling the mother liquor.

11. The spectral properties are as follows: IR (KBr disc) cm^{-1}: 2960 (s), 1450 (m), 1380 (s), 1315 (vs), 1205 (s), 1140 (vs), 1090 (vs), 745 (s); ^{1}H NMR (300 MHz, $CDCl_3$) δ: 1.56-1.82 (m, 4 H, C$\underline{H}_2$), 2.08-2.41 (m, 4 H, C$\underline{H}_2$), 2.15 (s, 3 H, C$\underline{H}_3$), 3.96 (dd, 1 H, C$\underline{H}SO_2$), 4.58 (AB quartet, 2 H, J_{AB} = 11, C$\underline{H}_2$Br); ^{13}C NMR ($CDCl_3$) δ: 22.66, 23.25, 24.27, 29.72, 43.69, 44.39, 65.81, 67.21.

12. Potassium tert-butoxide can be obtained from Aldrich Chemical Company, Inc.

13. The tert-butyl alcohol and tetrahydrofuran are distilled from calcium hydride prior to use.

14. Warming is required to dissolve the solid in this solvent.

15. The first four washings are done with gentle agitation to avoid emulsion formation.

16. The first cut of the distillate (ca. 1-2 mL) coming below 60°C was discarded.

17. The physical properties of the product are as follows: n_D^{20} 1.4722; IR (liquid film) cm^{-1}: 3090 (s), 2940 (s), 2870 (s), 1635 (s), 1440 (s), 895 (vs); ^{13}C NMR (CDCl$_3$) δ: 26.85, 35.37, 107.78, 149.68; GLC analysis (50 m OV-101 fused silica capillary column obtained from Perkin Elmer, Inc.) showed the product to be 90-93% pure. The material has an ^{1}H NMR spectrum corresponding to that reported in the literature:[4] (300 MHz, CDCl$_3$) δ: 1.62-1.66 (m, 4 H, C$\underline{H}_2$), 2.24-2.27 (m, 4 H, C$\underline{H}_2$), 4.64-4.65 (m, 2 H, vinyl C$\underline{H}$), 4.92-4.93 (m, 2 H, vinyl C$\underline{H}$).

3. Discussion

This procedure illustrates a recently published, simple, general method for the synthesis of conjugated dienes from olefins.[5] The scope of the reaction is shown in Table I.[5] In most of these examples hydrogen bromide elimination can be effected by stirring a solution of the olefin-bromomethanesulfonyl bromide adduct in methylene chloride with one equivalent of triethylamine at room temperature. Only two equivalents of the more costly potassium tert-butoxide are then needed in the second elimination step; the yields using the two-base procedure are generally superior to that obtained using only potassium tert-butoxide.

1,2-Dimethylenecyclohexane is a useful diene for Diels-Alder reactions[6,7,8] which has previously been synthesized in 11-77% yield in multistep procedures from cis-1,2-cyclohexanedicarboxylic anhydride[6] or acid,[7] from diethyl phthalate,[2] or from cyclohexanone.[4]

1. Department of Chemistry, State University of New York at Albany, Albany, NY 12222. This work was made possible by grants from the National Science Foundation, the Petroleum Research Fund administered by the American Chemical Society, the Société Nationale Elf Aquitaine and the John Simon Guggenheim Memorial Foundation.
2. Bailey, W. J; Golden, H. R. *J. Am. Chem. Soc.* **1953**, *75*, 4780; Blomquist, A. T.; Longone, D. T. *J. Am. Chem. Soc.* **1957**, *79*, 3916.
3. Bost, R. W.; Constable, E. W. *Org. Synth., Collect. Vol. III* **1943**, 610.
4. van Straten, J. W.; van Norden, J. J.; van Schaik, T. A. M.; Franke, G.Th.; de Wolf, W. H.; Bickelhaupt, F. *Recl. Trav. Chim. Pays-Bas* **1978**, *97*, 105.
5. Block, E.; Aslam, M. *J. Am. Chem. Soc.* **1983**, *105*, 6164; Block, E.; Aslam, M.; Eswarakrishnan, V.; Wall, A. *J. Am. Chem. Soc.* **1983**, *105*, 6165; Block E.; Aslam, M.; Iyer, R.; Hutchinson, J. *J. Org. Chem.* **1984**, *49*, 3664; Block, E.; Eswarakrishnan, V.; Gebreyes, K. *Tetrahedron Lett.* **1984**, *25*, 5469
6. Bartlett, P. D.; Wingrove, A. S.; Owyang, R. *J. Am. Chem. Soc.* **1968**, *90*, 6067.
7. Quin, L. D.; Leimert, J.; Middlemas, E. D.; Miller, R. W.; McPhail, A. T. *J. Org. Chem.* **1979**, *44*, 3496.
8. Thummel, R. P.; Cravey, W. E.; Nutakul, W. *J. Org. Chem.* **1978**, *43*, 2473.

Table I. Diene Synthesis via the Vinylogous Ramberg-Bäcklund Reaction

	Olefin	Product (Isomer Ratio[a])	Yield, %[b]
1	$C_4H_9CH=CH_2$	$C_3H_7CH=CHCH=CH_2$ (2:1)	38
2	$C_6H_{13}CH=CH_2$	$C_5H_{11}CH=CHCH=CH_2$ (2:1)[c]	61
3	(E)-$C_4H_9CH=CHC_4H_9$	(E)-$C_3H_7CH=CHC(C_4H_9)=CH_2$	68[d]
4	(E)-$C_5H_{11}CH=CHCH_3$	(E)-$C_4H_9CH=CHC(CH_3)=CH_2$	52
		+ $CH_2=CHC(C_5H_{11})=CH_2$	15
5	(cyclohexylidene-CH2)	(cyclohexenyl-vinyl)	53
6	(cycloheptylidene-CH2)	(cycloheptenyl-vinyl)	74
7	(cyclooctylidene-CH2)	(cyclooctenyl-vinyl)	75
8	(cyclohexene)	(methylenecyclohexene)	41[e]
9	(cycloheptene)	(methylenecycloheptene)	31[e]
10	(cyclooctene)	(methylenecyclooctene)	49[e]
11	(cycloheptene)	(methylenecycloheptene)	43[e]
12	$PhCH_2CH=CH_2$	$PhCH=CHCH=CH_2$ (1:8)	85
13	$PhOCH_2CH=CH_2$	$PhOCH=CHCH=CH_2$ (9:1)	54
14	$HO(CH_2)_9CH=CH_2$	$HO(CH_2)_8CH=CHCH=CH_2$ (5:1)	86
15	$(CH_3)_3SiCH_2CH=CH_2$	$(CH_3)_3SiCH=CHCH=CH_2$ (1:10)	41
16	$CH_2=CH(CH_2)_6CH=CH_2$	$CH_2=CH(CH_2)_5CH=CHCH=CH_2$	49[f,g,h]
17	$CH_2=CH(CH_2)_6CH=CH_2$	$CH_2=CHCH=CH(CH_2)_4CH=CHCH=CH_2$	40[i,j]
18	$(CH_3)_2Si(CH_2CH=CH_2)_2$	$(CH_3)_2Si(CH=CHCH=CH_2)_2$	38[i]

[a] (Z):(E) ratio. [b] Overall yield of distilled product. [c] (Z):(E) ratios 5:1 and 1:16 from (E)- and (Z)-1-octenylbromomethyl sulfones, respectively (59-61% overall distilled yields).[2] [d] Analysis by GLC indicated <1% (Z) isomer. [e] Et_3N step omitted. [f] Isomers not resolved by GLC. [g] Two equiv. of diene used. [h] Includes ca. 5% of 1,3,9,11-dodecatetraene. [i] Two molar equiv. of 1 used. [j] Ca. 80% (Z,Z).

Appendix

Chemical Abstracts Nomenclature (Collective Index Number); (Registry Number)

1,2-Dimethylenecyclohexane: Cyclohexane, 1,2-dimethylene (8); Cyclohexane, 1,2-bis(methylene)- (9); (2819-48-9)

Bromomethanesulfonyl bromide: Methanesulfonyl bromide, bromo- (9); (54730-18-6)

Sym-Trithiane: S-Trithiane (8); 1,3,5-Trithiane (9); (291-21-4)

Bromine (8,9); (506-68-3)

1-Methylcyclohexene: Cyclohexene, 1-methyl- (8,9); (591-49-1)

PREPARATION AND INVERSE ELECTRON DEMAND DIELS-ALDER REACTION OF AN ELECTRON-DEFICIENT DIENE: METHYL 2-OXO-5,6,7,8-TETRAHYDRO-2H-1-BENZOPYRAN-3-CARBOXYLATE

(2H-1-Benzopyran-3-carboxylic acid, 5,6,7,8-tetrahydro-2-oxo-, methyl ester)

A.

B.

Submitted by Dale L. Boger and Michael D. Mullican.[1]
Checked by Drew B. Burns and K. Barry Sharpless.

1. Procedure

A. Methyl 2-oxo-5,6,7,8-tetrahydro-2H-1-benzopyran-3-carboxylate. A dry, 500-mL, round-bottomed flask with a sidearm containing a magnetic stirring bar is fitted with a septum and a three-way stockcock equipped with an argon-filled balloon (Note 1). The air in the flask is replaced with argon (Note 2). Tetrahydrofuran (100 mL, Note 3) and diisopropylamine (4.7 g, 46 mmol, Note 4) are introduced into the flask through the septum using dry syringes (Note 5). The flask is immersed in an ice-water bath and a 2.8 M solution of butyllithium in hexane (17 mL, 46 mmol, Note 6) is added to the stirred solution using a syringe (10 min). The yellow solution is allowed to stir at 0°C for an additional 15 min. The resulting solution containing lithium diisopropylamide is immersed in a dry ice-2-propanol bath (-78°C) and a solution of cyclohexanone (3.74 g, 38.1 mmol, Note 7) in tetrahydrofuran (50 mL) is added using a syringe (30 min). The reaction is allowed to warm slowly to -5°C over 1.75 hr. The resulting solution containing the lithium enolate of cyclohexanone is recooled to -30°C to -25°C and a solution of dimethyl methoxymethylenemalonate (8.1 g, 46 mmol, Note 8) in tetrahydrofuran (20 mL) is added using a syringe (15 min). The reaction is allowed to warm to ambient temperature over 3.5 hr (Note 9). The reddish-orange solution is poured slowly onto aqueous 5% hydrochloric acid (300 mL) and the resulting yellow solution is extracted with methylene chloride (4 x 80 mL). The combined organic layers are dried over anhydrous sodium sulfate, filtered, and concentrated under reduced pressure to approximately 15 mL. The solution is applied to a medium pressure liquid chromatography column (25 x 500 mm, Note 10) packed with silica gel and 30% ethyl acetate-hexane. The eluant (30% ethyl acetate-hexane) is passed through the column at a rate of 20 mL/min; 20-mL fractions are collected (Note 11). The fractions are analyzed by thin-layer chromatography on analytical silica gel plates containing UV indicator (ethyl ether eluant). The fractions containing the product are combined and concentrated under reduced pressure to give 4.9 g (62%, 62-68%) of methyl 2-oxo-5,6,7,8-tetrahydro-2H-1-benzopyran-3-carboxylate as a white solid: mp 107-108°C (ethyl acetate-hexane, Note 12).

B. *6-Methoxy-7-methoxycarbonyl-1,2,3,4-tetrahydronaphthalene.* 1,1-Dimethoxyethylene (1.1 g, 12.5 mmol, Note 13) is added to a solution of methyl 2-oxo-5,6,7,8-tetrahydro-2H-1-benzopyran-3-carboxylate (507 mg, 2.44 mmol) in dry toluene (2.5 mL, Note 14) in a dry, 11 x 13 mm, resealable glass tube

(Note 15). The tube is flushed with argon and sealed with a Teflon plug. The reaction is warmed at 110°C in an oil bath for 15 hr (Note 16). The reaction is cooled and concentrated under reduced pressure. Purification of the product is effected by gravity chromatography on a 1.5 x 16-cm column of silica gel (30% ethyl ether-hexane eluant) collecting 5-mL fractions (Note 17). The fractions are analyzed by thin-layer chromatography (50% ethyl ether-hexane eluant) and those containing the product are combined and concentrated under reduced pressure to give 451 mg (84%, 84-86%) of 6-methoxy-7-methoxycarbonyl-1,2,3,4-tetrahydronaphthalene as a white solid: mp 98.5-99.5°C (methanol-water, Note 18).

2. Notes

1. The flask containing the stirring bar was dried at 120°C in an oven for several hours. The warm flask was fitted with a septum and a three-way stopcock.

2. This procedure is described in detail in *Org. Synth.* **1971**, *51*, 39.

3. Tetrahydrofuran was distilled from benzophenone ketyl under a nitrogen atmosphere.

4. Diisopropylamine was distilled from calcium hydride under a nitrogen atmosphere and stored over activated 3 Å sieve pellets.

5. The hypodermic syringes and needles were dried for several hours in an oven at 120°C and allowed to cool to ambient temperature in a desiccator.

6. Butyllithium was purchased from Aldrich Chemical Company, Inc.

7. Cyclohexanone was distilled before use.

8. Technical grade dimethyl methoxymethylenemalonate was purchased from Fluka Chemical Corporation and was purified by recrystallization from ether (2x), mp 43.0-44.0°C. It can be prepared by the procedure described for diethyl ethoxymethylenemalonate.[2]

9. Gradual warming to room temperature over 3.5 hr is necessary to ensure reasonable yields. Shorter times result in significantly lower yields.

10. The use of medium pressure liquid chromatography is described by Meyers.[3]

11. The checkers found that MPLC can be replaced by ordinary flash chromatography (30% EtOAc-hexane eluant, 6-cm I.D. column, ~ 200-240 g of flash-grade silica gel 230-400 mesh, 250-mL fractions). The crude product was dissolved in CH_2Cl_2 to which was added several grams of silica gel. This mixture was concentrated under reduced pressure and the resulting solid was applied to the top of the column.

12. The product has the following spectral properties: ^1H NMR ($CDCl_3$) δ: 1.80 (m, 4 H, C$\underline{H}_2$C$\underline{H}_2$), 2.46 (m, 4 H, CH_2C$\underline{H}_2$C=), 3.86 (s, 3 H, $-CO_2CH_3$), 7.99 (s, 1 H, vinyl); IR ($CHCl_3$) ν_{max} cm^{-1}: 3040, 2975, 1765, 1745, 1555, 1270, 1220, 1155.

13. 1,1-Dimethoxyethylene was purchased from Wiley Organics and used without further purification.

14. Toluene was distilled from calcium hydride under a nitrogen atmosphere.

15. The resealable glass tube was fabricated from a chromatography column purchased from Ace Glass Company. The tube was permanently sealed on one end and the other end remained internally threaded. A solid, threaded, Teflon plug equipped with an O-ring was used to seal the tube. Various sizes of such tubes are now available from Ace Glass Company.

16. *Caution: The reaction should be run behind a shield in a fume hood for protection in case of explosion. Pressure will build up in the tube since 1,1-dimethoxyethylene boils at 89°C and carbon dioxide is formed.*

17. The checkers found that gravity chromatography can be replaced by ordinary flash chromatography (30% ethyl ether-hexane eluant, 2.5-cm I.D. column ~ 40 g of flash grade silica gel, 20-mL fractions). In at least one case, the checkers found that pure product could be isolated in high yield (98%) without recrystallization.

18. The product has the following spectral properties: ^{1}H NMR (CDCl$_3$) δ: 1.80 (m, 4 H, C$\underline{H}_2$C$\underline{H}_2$), 2.75 (m, 4 H, CH$_2$C$\underline{H}_2$C=), 3.86 (s, 6 H, -OC$\underline{H}_3$ and -CO$_2$C$\underline{H}_3$), 6.65 (s, 1 H, C-5 H), 7.53 (s, 1 H, C-8 H); IR (CHCl$_3$) ν_{max} cm^{-1}: 3040, 2970, 1725, 1610, 1280, 1080, lit.[4] mp 99-100°C.

3. Discussion

This procedure describes the preparation and inverse electron demand (LUMO$_{diene}$ controlled)[5] Diels-Alder reaction of an electron-deficient diene. While extensive studies on the preparative utility of the normal (HOMO$_{diene}$ controlled)[5] Diels-Alder reaction have been detailed, few complementary studies on the preparative value of the inverse electron demand Diels-Alder reaction have been described.[6] Table I details representative 3-carbomethoxy-2-pyrones which have been prepared by procedures similar to that described herein and Tables II and III detail their inverse electron demand Diels-Alder reactions with electron-rich dienophiles.

A use of the LUMO$_{diene}$ controlled Diels-Alder reactions of 3-carbomethoxy-2-pyrones in the preparation of a full range of oxygenated aromatics [e.g., benzene, 1-, 2-, or 3-phenol, symmetrical and unsymmetrical o-catechol, resorcinol, and pyrogallol introduction (equation 1)][6f] as well as their application in the total synthesis of sendaverine, 6,7-benzomorphans, juncusol, imeluteine and rufescine, has been described.[6f]

$$\underset{1}{\text{R-CO-R'}} \longrightarrow \underset{2}{\text{pyranone-CO}_2\text{CH}_3} \longrightarrow \text{Ar(OR)}_n \quad (1)$$

n = 0-3

1. Department of Medicinal Chemistry, University of Kansas, Lawrence, KS 66045.
2. Fuson, R. C.; Parham, W. E.; Reed, L. J. *J. Org. Chem.* **1946**, *11*, 194.
3. Meyers, A. I.; Slade, J.; Smith, R. K.; Mihelich, E. D.; Hershenson, F. M.; Liang, C. D. *J. Org. Chem.* **1979**, *44*, 2247.
4. Arnold, R. T.; Zaugg, H. E.; Sprung, J. *J. Am. Chem. Soc.* **1941**, *63*, 1314
5. Houk, K. N. *J. Am. Chem. Soc.* **1973**, *95*, 4092.
6. (a) Ireland, R. E.; Anderson, R. C.; Badoud, R.; Fitzsimmons, B. J.; McGarvey, G. J.; Thaisrivongs, S.; Wilcox, C. S. *J. Am. Chem. Soc.* **1983**, *105*, 1988; (b) Bryson, T. A.; Donelson, D. M. *J. Org. Chem.* **1977**, *42*, 2930; (c) Corey, E. J.; Watt, D. S. *J. Am. Chem. Soc.* **1973**, *95*, 2303; (d) Markl, G.; Fuchs, R. *Tetrahedron Lett.* **1972**, 4695; (e) Behringer, H; Heckmaier, P. *Chem. Ber.* **1969**, *102*, 2835; (f) Boger, D. L.; Mullican, M. D. *Tetrahedron Lett.* **1982**, *23*, 4551, 4555; Boger, D. L.; Patel, M.; Mullican, M. D. *Tetrahedron Lett.* **1982**, *23*, 4559; Boger, D. L.; Mullican, M. D. *J. Org. Chem.* **1984**, *49*, 4033 and 4045; Boger, D. L.; Brotherton, C. E. *J. Org. Chem.* **1984**, *49*, 4050; and references cited therein.

Table I. Preparation of 3-Carbomethoxy-2-pyrones.[6f]

	Ketone	Method, % Yield	3-Carbomethoxy-2-pyrone	2
1a		A, 73%		2a
1b		A, 81%		2b
1c		B, 90%		2c
1d		B, 84%		2d
1e		B, 62%		2e
1f		B, 56%		2f
1g		B, 35% C, 59%		2g
1h		C, 47% D, 96%		2h
1i		D, 62%		2i

[a]Method A: The enolate of 1 was generated with 2.2 equiv of NaH in THF (0.2 M) at 0 to 25°C. Method B: the enolate of 1 was generated with 1.2 equiv of LDA in THF (0.2 M) at -78 to -5°C. Method C: the enolate of 1 was generated with 1.2 equiv of LDA in THF (0.2 M) at -78 to -5°C and closure to the α-pyrone was effected with acetic anhydride treatment at 100 to 130°C. Method D: the enolate of 1 was generated with 2.2 equiv of NaH in THF (0.2 M) at 0 to 25°C and closure to the α-pyrone was affected with catalytic p-toluenesulfonic acid treatment in refluxing toluene with distillative removal of methanol.

Table II. Diels-Alder Reaction of 3-Carbomethoxy-2-pyrones (2) with 1,1-Dimethoxyethylene: Salicylate Formation

3-Carbomethoxy-2-pyrone (2)	Conditions equiv time hr(temp. °C)	Product	% Yield[6f]
2a	8.5, 22(140)		59%
2b	10.0, 21(140)		75%
2c	10.0, 15(120)		78%
2d	5.5, 15(110)		86%
2e	5.0, 96(25), CH_2Cl_2 cat. $Ni(acac)_2$ 6.0, 12(95)		50% 90%
2f	10.0, 13(120)		80%
2g	10.0, 24(120)		91%
2i	8.0, 5(120)		83%

Table III. Diels Alder Reactions of 3-Carbomethoxy-2-pyrone (2d)[6f].

Entry	Dienophile (equiv.)	Conditions temp. °C (time, hr)	Product(s)	% Yield
1	N-vinyl pyrrolidinone (3.0)	160(42)	tetrahydronaphthalene-CO₂CH₃	98%
2	ethyl vinyl ether, OEt (5.0)	145(43)	" "	51%
3	vinylene carbonate (5-10.0)	180(40)	hydroxy-tetrahydronaphthalene-CO₂CH₃ (OH)	83%
4	1,1,2-trimethoxyethylene, OCH₃/OCH₃/OCH₃ (5.0)	150(78); cat. CH₃SO₃H or	tetrahydronaphthalene with OCH₃, OCH₃, CO₂CH₃	57%
	(10.0)	150(84); cat. DBU 150(12)	" "	51%
5	" " (10.0)	120(59)	tetrahydronaphthalene with CH₃O, OCH₃, OCH₃, CO₂CH₃	61%
6	H₃C-C≡C-NEt₂ (2.0)	150(17)	tetrahydronaphthalene with CH₃, NEt₂, CO₂CH₃	43%

106

Appendix

Chemical Abstracts Nomenclature (Collective Index Number); (Registry Number)

Methyl 2-oxo-5,6,7,8-tetrahydro-2H-1-benzopyran-3-carboxylate: 2H-1-Benzopyran-3-carboxylic acid, 5,6,7,8-tetrahydro-2-oxo-, methyl ester (11); (85531-80-2)

Diisopropylamine (8); 2-Propanamine, N-(1-methylethyl)- (9); (108-18-9)

Butyllithium: Lithium, butyl- (8,9); (109-72-8)

Cyclohexanone (8,9); (108-94-1)

Dimethyl methoxymethylenemalonate: Malonic acid, (methoxymethylene)-, dimethyl ester (8); Propanedioic acid, (methoxymethylene)-, dimethyl ester (9); (22398-14-7)

6-Methoxy-7-methoxycarbonyl-1,2,3,4-tetrahydronaphthalene: 2-Naphthalenecarboxylic acid, 5,6,7,8-tetrahydro-3-methoxy-, methyl ester (10); (78112-34-2)

1,1-Dimethoxyethylene: Ethene, 1,1-dimethoxy- (9); (922-69-0)

AMBIENT TEMPERATURE ULLMANN REACTION: 4,5,4',5'-TETRAMETHOXY-1,1'-BIPHENYL-2,2'-DICARBOXALDEHYDE

([1,1'-Biphenyl]-2,2'-dicarboxaldehyde, 4,4',5,5'-tetramethoxy-)

A. [structure with CH_3O, CHO, Br] $\xrightarrow[\text{reflux}]{C_6H_{11}NH_2,\ \text{toluene}}$ [imine product **1**]

B. $CuI + (C_2H_5O)_3P \longrightarrow CuI \cdot (C_2H_5O)_3P$ **2**

C. **1** $\xrightarrow[\text{2. } I_2]{\text{1. } C_4H_9Li,\ THF,\ -78°C}$ **3**

D. **1** $\xrightarrow{\begin{array}{l}\text{1. } C_4H_9Li,\ THF,\ -78°C\\ \text{2. 2, } -78°C\\ \text{3. 3, } -78°C\\ \text{4. } 25°C,\ 18\ hr\end{array}}$ **4**

Submitted by F. E. Ziegler, K. W. Fowler, W. B. Rodgers, and R. T. Wester.[1]
Checked by Tetsuji Oshima and Ryoji Noyori.

1. Procedure

CAUTION! Aqueous sodium cyanide is used in this procedure. All operations should be conducted in a well-ventilated hood and rubber gloves should be worn.

A. *6-Bromo-3,4-dimethoxybenzaldehyde cyclohexylimine* (**1**). A 2-L, three-necked flask is equipped with a Dean-Stark trap, reflux condenser, magnetic stirrer, and nitrogen inlet. The vessel is purged with nitrogen and charged with 40.0 g (0.16 mol) of 6-bromo-3,4-dimethoxybenzaldehyde (6-bromoveratraldehyde) (Note 1), 22.4 mL (0.20 mol) of cyclohexylamine (Note 2), and 800 mL of toluene. The mixture is refluxed for 16 hr (Note 3). The solution is cooled to room temperature and the solvent is removed on a rotary evaporator. The residual crystalline mass is recrystallized from a 3:1 hexane-methylene chloride mixture (1.5 L) to provide 48.4-51.4 g of the imine **1** as white crystals in two crops (mp 172-172.5°C)[2] (Note 4).

B. *Cuprous iodide-triethyl phosphite complex* (**2**). (Note 5). A 1-L, round-bottomed flask equipped with a magnetic stirrer and reflux condenser is flame-dried under nitrogen. The vessel is charged with 38.2 g (0.20 mol) of cuprous iodide, 33.4 mL (0.20 mol) of triethyl phosphite (Note 6) and 400 mL of dry toluene (distilled from CaH_2). The mixture is stirred at 80°C for 8 hr, cooled to room temperature, and filtered under reduced pressure on a pad of Celite. The solvent is removed from the filtrate successively with a rotary evaporator and briefly (15 min) under high vacuum. The residual solid is recrystallized from ether (25 mL) to yield 41.1-49.9 g (57.4-69.8%) of cuprous iodide-triethyl phosphite complex in two crops, mp 114-115°C.

C. *6-Iodo-3,4-dimethoxybenzaldehyde cyclohexylimine* (**3**). A 1-L, three-necked flask is equipped with a Claisen adapter (Note 7), Trubore stirrer, nitrogen inlet, and glass stoppers. The flask is thoroughly flame-dried under nitrogen. To the cooled flask is added 14.0 g (0.043 mol) of cyclohexylimine **1** and the glass stoppers are replaced by an alcohol thermometer and a rubber septum. Tetrahydrofuran (400 mL) (Note 8) is added via syringe through the septum and the resultant mixture is stirred at room temperature (27°C) for 30 min to effect solution. The solution is cooled to -78°C in a dry ice-acetone bath (Note 9). A solution of butyllithium in hexane (30.9 mL, 1.53 M, 0.047 mol) (Note 10) is added by syringe over 10 min at such a rate as to maintain the temperature below -75°C (Note 11). As the butyllithium is added the precipitate slowly dissolves leaving a clear, golden-yellow solution which is stirred for 15 min after the addition is complete. A solution of 27.0 g (0.11 mol) of iodine dissolved in 50 mL of dry THF is added via syringe to the reaction mixture at such a rate as to maintain the temperature below -70°C. The iodine solution (20-25 mL) is added until the red iodine color persists; precipitation also occurs. The mixture is warmed to room temperature, poured into 400 mL of water, and extracted with methylene chloride (5 x 400 mL). The combined organic extracts are dried (anhydrous $MgSO_4$), filtered, and concentrated on a rotary evaporator to 200 mL, and then washed with 200 mL of aqueous saturated sodium sulfite solution. The organic phase is redried, filtered, and concentrated. The solid residue is recrystallized from a 1:4 chloroform-hexane mixture (300 mL) to afford 12.6-13.5 g (78.6-85.0%) yield of 6-iodoveratraldehyde cyclohexylimine as white crystals, mp 180-181°C[2] (Note 4).

D. *4,5,4',5'-Tetramethoxy-1,1'-biphenyl-2,2'-carboxaldehyde* (**4**). The metalation procedure described in Section C is repeated using a 3-L flask, 18.4 g (0.056 mol) of 6-bromoveratraldehyde cyclohexylimine (**1**), 575 mL of tetrahydrofuran, and 40.6 mL (1.53 M, 0.062 mol) of butyllithium. After the metalation is complete at -78°C, the septum is replaced by a glass stopper. Solid cuprous iodide-triethyl phosphite complex (30.3 g, 0.085 mol) is added to the vessel at -78°C in one portion, immediately giving a green solution. The mixture is stirred for an additional 30 min. After the first 15 min, the solution turns a brownish-orange to red color. Solid 6-iodoveratraldehyde cyclohexylimine (**3**) (21.0 g, 0.056 mol) is added in one portion to produce an orange suspension. The reaction mixture is allowed to warm to room temperature (27°C) during which time the mixture becomes dark brown. The reaction mixture is stirred for 18 hr at room temperature. The reaction mixture is diluted with 600 mL of methylene chloride and 850 mL of 15% aqueous acetic acid and stirred vigorously for 17 hr. The yellow solution is transferred to a 4-L separatory funnel and the layers are separated. The organic layer is dried (anhydrous magnesium sulfate), filtered, and concentrated on a rotary evaporator to 800 mL and then transferred to a 2-L separatory funnel. The organic solution is washed with 5% aqueous hydrochloric acid (5 x 100 mL) and saturated aqueous sodium bicarbonate solution (10 x 50 mL). [*CAUTION: The final washings must be alkaline to avoid the liberation of hydrogen cyanide in the subsequent step.*] The organic layer is washed twice with 500 mL of 10% aqueous sodium cyanide solution, once with 500 mL of saturated aqueous sodium bicarbonate, and twice with 500 mL of water. [*CAUTION: The sodium cyanide washes should be bottled separately and labeled appropriately for approved disposal.*] The organic layer is dried (anhydrous magnesium sulfate), filtered, and concentrated to provide 16.3-20.3

g of residue. Crystallization from a 1:3 methylene chloride-hexane mixture at 5°C affords 13.1-16.9 g (70.4-90.7%) of beige crystals of the biphenyl, mp 215-216°C (lit[3] 214-215°C) after drying under high vacuum (0.1 mm) (Note 4).

2. Notes

1. 6-Bromo-3,4-dimethoxybenzaldehyde was purchased from the Aldrich Chemical Company, Inc. (Milwaukee) or readily prepared by bromination of veratraldehyde (Aldrich Chemical Company, Inc. or Tokyo Kasei).[4]

2. Cyclohexylamine (Aldrich Chemical Company, Inc. or Nakarai Chemicals) and all solvents and reagents (reagent grade) were used as received, unless otherwise specified.

3. The water level in the trap remains constant after this period of time.

4. Spectral characterization; ^{1}H NMR (CDCl$_3$). 6-Bromoveratraldehyde cyclohexylimine, δ: 1.07-1.82 (m, 10 H), 3.30 (m, 1 H, NC$\underline{H}$), 3.89 (s, 3 H, OC$\underline{H}_3$) 3.92 (s, 3 H, OC$\underline{H}_3$), 6.97 (s, 1 H), 7.55 (s, 1 H), and 8.54 (s, 1 H, N=C$\underline{H}$); 6-Iodoveratraldehyde cyclohexylimine, δ: 1.07-1.82 (m, 10 H), 3.31 (m, 1 H, NC$\underline{H}$), 3.88 (s, 3 H, OC$\underline{H}_3$), 3.92 (s, 3 H, OC$\underline{H}_3$), 7.22 (s, 1 H), 7.53 (s, 1 H), and 8.32 (s, 1 H, N=C$\underline{H}$); 4,5,4',5'-Tetramethoxy-1,1'-biphenyl-2,2'-dicarboxaldehyde, δ: 3.96 (s, 6 H, OC$\underline{H}_3$), 4.01 (s, 6 H, OC$\underline{H}_3$), 6.80 (s, 2 H), 7.56 (s, 2 H), and 9.67 (s, 2 H, C$\underline{H}$O).

5. This method was adapted from the procedure of Nishizawa.[5] The complex is reported to have mp 109-110°C.[6]

6. Cuprous iodide was purchased from Alfa Products, Morton/Thiokol Inc. or Kishida Chemicals and triethyl phosphite from the Aldrich Chemical Company, Inc. or Nakarai Chemicals.

7. The offset neck of the adapter was fitted with the nitrogen inlet and the other neck with a glass stopper which was eventually replaced with a thermometer.

8. Tetrahydrofuran (THF) was distilled from sodium benzophenone ketyl under nitrogen in all applications.

9. The bromide, 1, precipitated during the cooling.

10. Butyllithium was purchased from Alfa Products, Morton/Thiokol Inc. or Mitsuwa Pure Chemicals and was standardized by the method of Kofron.[7]

11. Alcohol thermometers were found not to read temperatures accurately. A temperature of -78°C designates the lowest temperature to which a large dry ice-acetone bath cools the reaction mixture. The temperature -75°C signifies a 3° rise in temperature.

3. Discussion

The Ullmann reaction has been traditionally conducted at elevated temperatures (100-250°C), with or without solvent, in the presence of copper powder. Often the quality of copper can be extremely important to the success of the reaction.[8] Aromatic bromides and halides which bear ortho-substituted electron-withdrawing groups undergo coupling at the low end of the temperature range. Cross coupling is best accomplished when only one of the aryl halides bears an electron-withdrawing group.[9] In such instances, an excess of the aryl halide without the electron-withdrawing group may have to be employed.[10]

Nickel(0) reagents have been employed in the symmetrical coupling of aryl halides in sterically unencumbered cases.[11,12] An efficient cross coupling reaction between an arylzinc halide and an ortho-iodoarylimine under mild conditions mediated by Ni(0) has been reported.[13] Thallium(III) trifluoroacetate has been employed in the symmetrical coupling of aromatic ethers.[14] The use of diazonium salts in the formation of unsymmetrical biphenyls has been reviewed.[15]

The present method permits both symmetrical and unsymmetrical coupling to occur at room temperature. It is necessary for a substitutent (nitrogen or sulfur) to be situated ortho to the halogen so that the heteroatom can chelate well with copper. This requirement must be fulfilled in both reacting partners. The organolithium species may be generated by metal-hydrogen or metal-halogen exchange. The coupling works well in sterically congested compounds, and only for aryl iodides. o-Iodoaldehydes may also be prepared by direct iodination of aromatic aldehydes.[12,16] Representative applications of this reaction are provided in the Table.

Table I. Ambient Temperature Ullman Reaction

Organocopper	Iodide	Biphenyl	Yield (%)[ref]
			57[2]
			44[2]
			76(62)[2]
			63[2]
			88[12]

Table I. (continued) Ambient Temperature Ullman Reaction

Organocopper	Iodide	Biphenyl	Yield (%)[ref]
			54[2]
			48[2]
			63[2]

1. Department of Chemistry, Yale University, New Haven, CT 06511.
2. Ziegler, F. E.; Chliwner, I.; Fowler, K. W.; Kanfer, S. J.; Kuo, S. J.; Sinha, N. D. *J. Am. Chem. Soc.* **1980**, *102*, 790.
3. Bick, I. R. C.; Harley-Mason, J.; Sheppard, N.; Vernengo, M. J. *J. Chem. Soc.* **1961**, 1896.
4. Pschorr, R., *Justus Liebigs Ann. Chem.* **1912**, *391*, 23.
5. Nishizawa, Y. *Bull. Chem. Soc. Jpn.* **1961**, *34*, 1170
6. Arbusott, A, *Ber.* **1905**, *38*, 1171
7. Kofron, W. G.; Baclawski, L. M. *J. Org. Chem.* **1976**, *41*, 1879.
8. Newman, M. S.; Dali, H. M. *J. Org. Chem.* **1977**, *42*, 734.
9. Ikeda, T.; Taylor, W. I.; Tsuda, Y.; Uyeo, S.; Yajima, H. *J. Chem. Soc.* **1956**, 4749; Jeffs, P. W.; Hansen, J. F.; Brine, G. A. *J. Org. Chem.* **1975**, *40*, 2883; Koizumi, J.; Kobayashi, S.; Uyeo, S. *Chem. Pharm. Bull.* **1964**, *12*, 696.
10. Brown, E.; Robin, J.-P. *Tetrahedron Lett.* **1977**, 2015.
11. Semmelhack, M. F.; Helquist, P. M.; Jones, L. D. *J. Am. Chem. Soc.* **1971**, *93*, 5908; Kende, A. S.; Liebeskind, L. S.; Braitsch, D. M. *Tetrahedron Lett.* **1975**, 3375.
12. Kende, A. S.; Curran, D. P. *J. Am. Chem. Soc.* **1979**, *101*, 1857.
13. Larson, E. R.; Raphael, R. A. *Tetrahedron Lett.* **1979**, 5041.
14. McKillop, A.; Turrell, A. G.; Young, D. W.; Taylor, E. C. *J. Am. Chem. Soc.* **1980**, *102*, 6504.
15. Bachmann, W. E.; Hoffman, R. A. "Organic Reactions", Wiley: New York, 1944; Vol. II, p. 224.
16. Janssen, D. E.; Wilson, C. V. *Org. Synth., Collect. Vol. IV* **1963**, 547.

Appendix

Chemical Abstracts Nomenclature (Collective Index Number); (Registry Number)

4,5,4',5'-Tetramethoxy-1,1'-biphenyl-2,2'-dicarboxaldehyde: [1,1'-Biphenyl]-2,2'-dicarboxaldehyde, 4,4',5,5'-tetramethoxy- (9); (29237-14-7)

6-Bromo-3,4-dimethoxybenzaldehyde cyclohexylimine: Cyclohexanamine, N-[(2-bromo-4,5-dimethoxyphenyl)methylene]- (10); (73252-55-8)

6-Bromo-3,4-dimethoxybenzaldehyde: Benzaldehyde, 2-bromo-4,5-dimethoxy- (9); (5392-10-9)

Cyclohexylamine (8); Cyclohexanamine (9); (108-91-8)

Cuprous iodide: Copper iodide (8,9); (7681-65-4)

Triethyl phosphite: Phosphorous acid, triethyl ester (8,9); (122-52-1)

6-Iodo-3,4-dimethoxybenzaldehyde cyclohexylimine: Cyclohexanamine, N-[(2-iodo-4,5-dimethoxyphenyl)methylene]- (10); (61599-78-8)

Butyllithium: Lithium, butyl- (8,9); (109-72-8)

Iodine (8,9); (7553-56-2)

GEMINAL ACYLATION-ALKYLATION AT A CARBONYL CENTER USING DIETHYL N-BENZYLIDENEAMINOMETHYLPHOSPHONATE: PREPARATION OF 2-METHYL-2-PHENYL-4-PENTENAL

(4-Pentenal, 2-methyl-2-phenyl-)

A. Phthalimide-N-CH$_2$OH (**1**) $\xrightarrow[\text{60-70°C}]{\text{48% aq HBr, concd H}_2\text{SO}_4}$ Phthalimide-N-CH$_2$Br (**2**)

B. **2** $\xrightarrow[\text{85-100°C}]{\text{(EtO)}_3\text{P}}$ Phthalimide-N-CH$_2$-P(OEt)$_2$=O (**3**) $\xrightarrow[\text{EtOH/RT} \to \Delta]{\text{95% N}_2\text{H}_4}$ (EtO)$_2$P-CH$_2$NH$_2$ (**4**)

C. **4** $\xrightarrow[\text{RT} \to \Delta]{\text{PhCHO/C}_6\text{H}_6}$ (EtO)$_2$P-CH$_2$N=CHPh (**5**) $\xrightarrow[\text{2) PhCOCH}_3\text{/-78°C} \to \Delta]{\text{1) BuLi/THF/-78°C}}$ Ph(Me)C=CH-N=CHPh (**6**)

6 $\xrightarrow[\text{3) H}_3\text{O}^+\text{/RT}]{\substack{\text{1) BuLi/-78°C} \\ \text{2) CH}_2\text{=CHCH}_2\text{Br/-78°C} \to \text{RT}}}$ Ph(Me)C(CHO)CH$_2$CH=CH$_2$ (**7**)

Submitted by Steven K. Davidsen, Gerald W. Phillips, and Stephen F. Martin.[1]

Checked by Robert G. Aslanian, Max Tishler, and Andrew S. Kende.

119

1. Procedure

A. *N-Bromomethylphthalimide* (2). A 1-L, three-necked, round-bottomed flask is fitted with a mechanical stirrer, a 125-mL pressure equalizing dropping funnel, and a thermometer. The flask is charged with 50.0 g (0.28 mol) of N-hydroxymethylphthalimide (Note 1) and 200 mL of 48% aqueous hydrobromic acid (Note 2). The flask is immersed in an ice bath, and 75 mL of concentrated sulfuric acid is added with stirring over a period of about 15 min (Note 3). Upon completion of the addition, the flask is removed from the ice bath, heated at 60-70°C for 5 hr and then cooled overnight in a refrigerator. The solid is collected by suction filtration using a 125-mm glass funnel with a coarse frit. The crude product is washed thoroughly with three 100-mL portions of cold water, two 50-mL portions of cold 10% aqueous ammonium hydroxide and finally with three 100-mL portions of cold water (Note 4). The crude product thus obtained is completely dried under reduced pressure at room temperature over phosphorus pentoxide to give 57.1-63.8 g (85-95%) of N-bromomethylphthalimide as a light tan solid, mp 142-147°C. Although the material thus obtained may be used in the next step without further purification, it may also be recrystallized from dry acetone, mp 147-148°C (lit., mp 148°C,[2] 148-149°C[3]) (Note 5).

B. *Diethyl phthalimidomethylphosphonate* (3). A 500-mL, one-necked, round-bottomed flask equipped with a magnetic stirring bar and an efficient reflux condenser (approximately 80-cm long) is charged with 51.2 g (0.21 mol) of dry N-bromophthalimide (2) and 43.1 g (0.26 mol) of freshly distilled triethyl phosphite (Note 6). The mixture is immersed in an oil bath and the temperature of the oil bath gradually increased over about 15 min to 85-100°C, whereupon the solid dissolves and a vigorous, exothermic reaction ensues

(Note 7). When the reaction has subsided, the oil bath is lowered and the condenser is removed. The flask is fitted for simple distillation, and ethyl bromide is distilled from the reaction mixture over a period of 3-4 hr by continued heating at 115°C (oil bath). The resulting light yellow oil is cooled to room temperature, whereupon it solidifies (Note 8). The crude product is collected by suction filtration and washed with three 100-mL portions of cold petroleum ether (bp 60-68°C) to give 50.1-56.3 g (80-90%) of diethyl phthalimidomethylphosphonate as white crystals, mp 60-63°C, which are used in the next step without further purification. Recrystallization of this material from diethyl ether/petroleum ether (bp 60-68°C) affords pure 3, mp 65-67°C (lit.,[3] 67°C) (Note 9).

C. *Diethyl N-benzylideneaminomethylphosphonate* (5). A 2-L, two-necked, round-bottomed flask is equipped with a mechanical stirrer and a 125-mL pressure equalizing dropping funnel fitted with a calcium chloride drying tube. The flask is charged with 50.0 g (0.17 mol) of diethyl phthalimidomethylphosphonate (3) dissolved in 750 mL of absolute ethanol (Note 10). To this solution is then added 6.4 g (0.19 mol) of 95% hydrazine (Note 11) in 50 mL of absolute ethanol, and the resulting mixture is stirred overnight at room temperature. The dropping funnel is replaced with a reflux condenser bearing a calcium chloride drying tube, and the mixture is heated at reflux for 4 hr and then cooled to 0-5°C in an ice bath. The precipitated phthalhydrazide is collected by suction filtration and thoroughly washed with three 125-mL portions of benzene. The excess solvents and hydrazine are completely removed under reduced pressure on a rotary evaporator and then under high vacuum (Notes 12, 13). The crude diethyl aminomethylphosphonate (**4**)[4,5] (Note 14) thus obtained is dissolved in 350 mL of reagent grade benzene, and the solution is cooled overnight in the refrigerator. Any additional phthalhydrazide which precipitates is removed by suction filtration and washed with two 25-mL portions of benzene. The filtrate and washings are combined in a 1-L, one-necked flask equipped with magnetic stirring bar, and the solution is cooled to 5-10°C at which time 21.2 g (0.20 mol) of freshly distilled benzaldehyde is added in one portion with stirring. The mixture is stirred for 4 hr at room temperature, the flask is fitted with a Dean-Stark trap and a reflux condenser and heated overnight at reflux with constant removal of water. The solution is cooled to approximately room temperature, and the excess solvents are removed under reduced pressure. The crude product is purified by vacuum distillation, bp 145-149°C (0.05 mm), to give 37.5-40.2 g (84-90%) of pure diethyl N-benzylideneaminomethylphosphonate (5)[6,7] as a light yellow oil (Note 15).

D. *2-Methyl-2-phenyl-4-pentenal* (7). A dry, 100-mL three-necked, round-bottomed flask with 14/20 joints is fitted with a magnetic stirrer, reflux condenser, and a rubber septum (Note 16). The flask is charged with 50 mL of anhydrous tetrahydrofuran (Note 17) and cooled to -78°C in a dry ice/isopropyl alcohol bath, and a solution of butyllithium (12.0 mmol) in hexane (Note 18) is added with stirring. To this stirred solution is added dropwise via syringe a solution of 3.06 g (12.0 mmol) of diethyl N-benzylideneamino-methylphosphonate (5) in 5 mL of anhydrous tetrahydrofuran, and the colored solution is stirred an additional hour at -78°C. A solution containing 1.20 g (10.0 mmol) of freshly distilled acetophenone in 5 mL of anhydrous tetrahydrofuran is added dropwise, and the cooling bath is removed. The solution is stirred for 1 hr at room temperature and then at reflux for 2 hr. After the solution is cooled to room temperature, it is poured into a 250-mL round-bottomed flask and the solvents are removed under reduced pressure on a rotary evaporator. The yellow residue is partitioned between

50 mL of ether and 50 mL of saturated sodium chloride. The layers are separated and the aqueous phase is extracted with three 25-mL portions of ether. The combined organic layers are washed with 50 mL of saturated sodium chloride and dried over magnesium sulfate. Magnesium sulfate is removed by filtration, and the excess solvents are then completely removed under reduced pressure on a rotary evaporator. The resulting yellow solid is dried under reduced pressure and transferred to a 100-mL two-necked, round-bottomed flask which is fitted with a magnetic stirring bar, a nitrogen inlet, and a rubber septum. The flask is charged with 50 mL of anhydrous tetrahydrofuran and flushed thoroughly with dry nitrogen. The resulting solution of the 2-azadiene 6 (Note 20) is cooled to -78°C, and a solution of butyllithium (12.0 mmol) in hexane is added dropwise via syringe. The deeply-colored solution is stirred at -78°C for 1 hr at which time 1.81 g (15.0 mmol) of freshly distilled allyl bromide (Note 21) is added. The cooling bath is removed, and the solution is stirred for 2 hr at room temperature. The reaction is added to 50 mL of 3 N aqueous hydrochloric acid, and the resulting heterogeneous mixture is stirred vigorously for 18 hr at room temperature. After the addition of 50 mL of saturated sodium chloride, the layers are separated, and the aqueous layer is extracted with three 75-mL portions of ether. The combined organic layers are washed with 75-mL portions of saturated aqueous sodium bicarbonate and saturated sodium chloride, and the washings are backwashed with a 50-mL portion of ether. The combined organic layers are dried over magnesium sulfate, and the excess solvents are removed under reduced pressure on a rotary evaporator. Distillation of the resulting yellow oil under reduced pressure gives 1.30-1.45 g (75-83%) of pure 2-methyl-2-phenyl-4-pentenal as a colorless liquid, bp 70-73°C (0.1 mm) (Note 22).

2. Notes

1. Although N-hydroxymethylphthalimide may be purchased from Aldrich Chemical Company, Inc., it may also be prepared from phthalimide and 37% aqueous formaldehyde.[8] Material prepared in this way should be dried at room temperature under reduced pressure over phosphorus pentoxide.

2. Aqueous 48% hydrobromic acid should be purchased from Eastman Organic Chemicals since that obtained from other sources tends to give, for unknown reasons, less satisfactory results.

3. The temperature should not be allowed to exceed 30°C during the addition.

4. Removal of *all* of the hydrobromic acid by washing is critical to the success of the next reaction. If the filtrate is not basic after washing with cold 10% aqueous ammonium hydroxide the washing should be continued until the filtrate is basic. Disconnection of the vacuum during each washing is recommended. The final aqueous wash should be no more basic than pH 8-9. Use of a rubber dam facilitates the filtration and washing.

5. The proton magnetic resonance spectrum of 2 exhibits the following absorptions (CDCl$_3$) δ: 5.42 (s, 2 H, C$\underline{\text{H}}_2$Br), 7.65-7.91 (complex, 4 H, aromatic).

6. Triethyl phosphite was purchased from Aldrich Chemical Company, Inc.

7. The exothermic reaction usually commences after all of the solid has dissolved. It is important to allow this exothermic reaction, which lasts about 5 min, to run its course without cooling, since premature cooling results in lower yields of impure product which may be difficult to purify.

8. Use of impure N-bromomethylphthalimide or incomplete reaction may lead to the formation of a gummy or mushy residue at this stage, and the addition of petroleum ether (bp 60-68°C) might facilitate crystallization. Alternatively, the crude product may be purified by dissolving it in a minimum volume of anhydrous ether, addition of petroleum ether (bp 60-68°C) until the solution turns cloudy, and then cooling. Scratching may be necessary to induce crystallization. Several crops of crystals may be collected, but the total yields thus obtained are generally lower than 80%.

9. The proton magnetic resonance spectrum of product 3 exhibits the following absorptions (CDCl$_3$) δ: 1.31 (t, 6 H, J = 7, CH$_2$C$\underline{H}_3$), 3.83-4.26 (complex, 6 H, OC$\underline{H}_2$, NC$\underline{H}_2$P), 7.60-7.86 (complex, 4 H, aromatic).

10. Absolute ethanol was purchased from Aaper Alcohol and Chemical Company and used without further purification. Slight heating may be required to effect solution.

11. Hydrazine, 95%, was purchased from Eastman Organic Chemicals.

12. Since it may undergo reaction with benzaldehyde in the subsequent step to give benzaldehyde azine, it is advisable to remove the last traces of hydrazine by rotating the flask under reduced pressure. The submitters used an oscillating motor which operates on compressed air or vacuum and is commonly employed with Kugelrohr distilling units. One such motor is available from the Aldrich Chemical Company, Inc.

13. Diethyl aminomethylphosphonate undergoes decomposition upon attempted distillation, but no deterioration of the product was observed if these operations were executed at temperatures not exceeding 30°C.

14. The proton magnetic resonance spectrum of crude **4** exhibits the following absorptions (CDCl$_3$) δ: 1.31 (t, 6 H, J = 7, CH$_2$C$\underline{H}_3$), 2.68 (br s, 2 H, N$\underline{H}_2$), 3.00 (d, 2 H, J = 10, PC$\underline{H}_2$N), 3.95-4.27 (complex, 4 H, aromatic).

15. The proton magnetic resonance spectrum of **5** exhibits the following absorptions (CDCl$_3$) δ: 1.32 (t, 6 H, J = 7, CH$_2$C$\underline{H}_3$), 3.85-4.24 (complex, 6 H, C$\underline{H}_2$CH$_3$, PC$\underline{H}_2$N), 7.24-7.40 (complex, 3 H, para, meta Ph CH), 7.64-7.75 (complex, 2 H, ortho Ph CH), 8.25 (d, 1 H, J = 5, N=C$\underline{H}$Ph). This material shows no significant tendency to deteriorate when stored under dry nitrogen at room temperature.

16. The apparatus was flame dried under a flow of dry nitrogen and then kept under a slight positive pressure of nitrogen during the reactions by maintaining a slow flow of nitrogen through a mercury bubbler.

17. Tetrahydrofuran was distilled from the potassium ketyl of benzophenone. *(Caution:* See *Org. Synth., Collect. Vol. 5* **1973**, 976 for a warning regarding the purification of tetrahydrofuran.)

18. Butyllithium was prepared by dilution of 90% butyllithium obtained from Lithium Corporation of America with purified hexane or petroleum ether (bp 60-68°C) (Note 19). The normality was determined prior to use by titration according to the method of Watson and Eastman.[9]

19. Hexane was purified by preliminary stirring over concentrated sulfuric acid and then anhydrous potassium carbonate followed by distillation. The hexane thus obtained was then distilled from sodium wire.

20. The proton magnetic resonance spectrum of **6** exhibits the following absorptions (CDCl$_3$) δ: 2.52 (br s, 3 H, C-C$\underline{H}_3$), 7.20-7.65 (complex, 9 H, aromatic), 7.86 (m, 2 H, aromatic), 8.33 (br s, 1 H, N=C$\underline{H}$Ph).

21. Allyl bromide was purchased from Aldrich Chemical Company, Inc., and distilled from pulverized calcium hydride and filtered through basic alumina (10 g) immediately prior to use.

22. The proton magnetic resonance spectrum of 7 exhibits the following absorptions (CDCl$_3$) δ: 1.32 (s, 3 H, CC$\underline{H}_3$), 2.49 (d, 2 H, J = 7, C$\underline{H}_2$CH=CH$_2$), 4.88 (m, 2 H, CH=C$\underline{H}_2$), 5.45 (m, 1 H, C$\underline{H}$=CH$_2$), 7.14 (m, 5 H, aromatic), 9.33 (s, 1 H, C$\underline{H}$O).

3. Discussion

The procedure in the present reaction sequence for the preparation of N-bromomethylphthalimide (2) is a modification of that reported by Pucher and Johnson.[2] N-Bromomethylphthalimide has also been prepared by treatment of N-hydroxymethylphthalimide with phosphorus tribromide.[3] The procedures for the syntheses of the phosphonates 3 and 4 represent modifications of those described by Yamauchi and co-workers.[3,4] Two other routes to 4 have recently been reported by Gross and co-workers.[5] Ratcliffe and Christensen have also recorded the preparation of diethyl N-benzylideneaminomethylphosphonate (5) by the condensation of benzaldehyde with 4 under virtually identical conditions as detailed herein, but their route to 4 was completely different.[6] The submitters have found that the present method for the syntheses of 2-5 gives reproducibly higher yields and is more reliable and convenient than those alternative procedures.

Diethyl N-benzylideneaminomethylphosphonate (5) has been previously employed as an intermediate in the synthesis of β-lactam antibiotics[6] and as a reagent for the homologation of aldehydes and ketones via intermediate 2-azadienes.[7] Other derivatives of dialkyl aminomethylphosphonates have also emerged as useful synthetic reagents. For example, diethyl isocyanomethylphosphonate (8) may be employed for the conversion of aldehydes and ketones to

$(EtO)_2PCH_2NC$
8

$(MeO)_2PCHN_2$
9

α,β-unsaturated isocyano compounds.[10] Dimethyl diazomethylphosphonate (9)[11] has recently been shown to be an effective reagent for the elaboration of aldehydes or alkyl aryl ketones and diaryl ketones to alkynes[12] and for the conversion of dialkyl ketones into aldehydic enol ethers and enamines.[13]

Part D of the present procedure represents a slight modification of a general method for effecting the net replacement of both of the carbon-oxygen bonds of a carbonyl group with an acyl group and a functionalized alkyl appendage; some typical examples of the original procedure are collected in Table I.[14] Moreover, when the electrophile employed for alkylation of the intermediate metalloenamine[15] is properly selected, it is possible to introduce geminal substituents at the carbonyl function that are suitably functionalized for subsequent conversion either to 4,4-disubstituted cyclopentenones (eq. 1)[14] or 4,4-disubstituted cyclohexenones (eq. 2).[15]

a. $(EtO)_2P(O)CHLiN=CHPh$/THF/-78°C →reflux; b. BuLi/-78°C;
c. $CH_2=CClCH_2Cl$/THF/HMPA/-78°C →reflux; d. H_3O^+; e. $Hg(OAc)_2$/HCO_2H/RT
f. KOH/aq MeOH/RT

The preparation of **10** represents a formal total synthesis of α-cuparenone. An annulation related to that depicted in eq. 2, which was a key step in an

$$\text{(2)}$$

a. $(EtO)_2P(O)CHLiN=CHPh/THF/-78°C \rightarrow$ reflux; b. BuLi/-78°C;
c. $BrCH_2CH_2C(OCH_2CH_2O)CH_3/THF/HMPA/-78°C \rightarrow RT$;
d. H_3O^+/RT; e. KOH/aq MeOH/RT

efficient synthesis of mesembrine,[16] has also been featured in total syntheses of the Amaryllidaceae alkaloids lycoramine[17] and crinine.[18] Finally, the intermediate metalloenamines may be utilized as the nucleophilic partners in directed aldol reactions (eq. 3), but it appears to be necessary to trap the intermediate β-oxidoimines by acylation or alkylation to avoid retro-aldolization during the hydrolysis step.[14] Such a process has recently been exploited in the syntheses of the Amaryllidaceae alkaloids pretazettine and haemanthidine.[19]

$$\text{(3)}$$

a. $(EtO)_2P(O)CHLiN=CHPh/THF/-78°C \rightarrow$ reflux; b. BuLi/-78°C;
c. PhCHO; d. MeOCOCl/-78°C $\rightarrow$RT; e. H_3O^+/RT

It presently appears that this methodology is well suited for the construction of quaternary carbon atoms bearing substituted alkyl appendages containing a diverse array of functionality. In large measure this is because metalloenamines, which are the key synthetic intermediates, are highly nucleophilic and generally undergo regioselective reaction at carbon with a variety of weak and multifunctional electrophiles. Moreover, numerous functional groups may be present on the starting aldehyde or ketone, but there has been a report[20] which suggests that carbonyl compounds bearing potential leaving groups on the carbon adjacent to the carbonyl group may not be good substrates. While a number of individual operations are required, it is frequently possible to execute the entire sequence of reactions in a single flask. In this particular preparation the sequence may also be performed in a single vessel, but purification of product 7 by simple distillation is more difficult because of the presence of lower boiling impurities.

Table I. Geminal Acylation-Alkylation of Carbonyl Compounds

Entry	R_1	R_2	R_3	% Overall Yield[a]
1	C_6H_{13}	H	Me	34
2	c-C_6H_{11}	H	Me	43
3	n-Pr	n-Pr	Me	58
4	-$(CH_2)_5$-		$CH_2CH=CH_2$	51
5	-$(CH_2)_2CH(t-Bu)(CH_2)_2$-		Me	63[b]
6	-$CH_2C(Me)_2CH_2C(Me)=CH$-		Me	80
7	Ph	Me	Me	77

a. Yields are of distilled products but are not optimized.
b. As a ca. 79:21 mixture of diastereomers.

1. Department of Chemistry, The University of Texas at Austin, Austin, TX 78712.
2. Pucher, G. W.; Johnson, T. B. *J. Am. Chem. Soc.* **1922**, *44*, 817.
3. Yamauchi, K.; Kinoshita, M.; Imoto, M. *Bull. Chem. Soc. Jap.* **1972**, *45*, 2531.
4. Yamauchi, K.; Mitsuda, Y.; Kinoshita, M. *Bull. Chem. Soc. Jap.* **1975**, *48*, 3285.
5. Gross, H.; Beisert, S.; Costisella, B. *J. Prakt. Chem.* **1981**, *323*, 877.
6. Ratcliffe, R. W.; Christensen, B. G. *Tetrahedron Lett.* **1973**, 4645.
7. Dehnel, A.; Finet, J. P.; Lavielle, G. *Synthesis* **1977**, 474.
8. Buc, S. R. *J. Am. Chem. Soc.* **1947**, *69*, 254.
9. Watson, S. C.; Eastman, J. F. *J. Organomet. Chem.* **1967**, *9*, 165.
10. Schöllkopf, U.; Schröder, R.; Stafforst, D. *Justus Liebigs Ann. Chem.* **1974**, 44; Schöllkopf, U.; Eilers, E.; Hantke, K. *Justus Liebigs Ann. Chem.* **1976**, 969.
11. Seyferth, D.; Marmor, R. S.; Hilbert, P. *J. Org. Chem.* **1971**, *36*, 1379.
12. Gilbert, J. C.; Weerasooriya, U. *J. Org. Chem.* **1982**, *47*, 1837.
13. Gilbert, J. C.; Weerasooriya, U. *Tetrahedron Lett.* **1980**, *21*, 2041.; Gilbert, J. C.; Weerasooriya, U.; Wiechman, B.; Ho, L. *Tetrahedron Lett.* **1980**, *21*, 5003.
14. Martin, S. F.; Phillips, G. W.; Puckette, T. A.; Colapret, J. A. *J. Am. Chem. Soc.* **1980**, *102*, 5866.
15. For another example of the generation of metalloenamines from 2-azadienes see Wender, P. A.; Eissenstat, M. A. *J. Am. Chem. Soc.* **1978**, *100*, 292; Wender, P. A.; Schaus, J. M. *J. Org. Chem.* **1978**, *43*, 782.
16. Martin, S. F.; Puckette, T. A.; Colapret, J. A. *J. Org. Chem.* **1979**, *44*, 3391.

17. Martin, S. F.; Garrison, P. J. *J. Org. Chem.* **1981**, *46*, 3567; Martin, S. F.; Garrison, P. J. *J. Org. Chem.* **1982**, *47*, 1513.
18. Martin, S. F., Campbell, C. L., unpublished results.
19. Martin, S. F.; Davidsen, S. K. *J. Am. Chem. Soc.* **1984**, *106*, 6431.
20. For example see: Jarosz, S.; Fraser-Reid, B. *Tetrahedron Lett.* **1981**, *22*, 2533.

Appendix
Chemical Abstracts Nomenclature (Collective Index Number); (Registry Number)

Diethyl N-benzylideneaminomethylphosphonate: Phosphonic acid, [[(phenylmethylene)amino]methyl]-, diethyl ester (9); (50917-73-2)

2-Methyl-2-phenyl-4-pentenal: 4-Pentenal, 2-methyl-2-phenyl- (8,9); (24401-39-6)

N-Bromomethylphthalimide: Phthalimide, N-(bromomethyl)- (8); 1H-Isoindole-1,3-(2H)-dione, 2-(bromomethyl)- (9); (5332-26-3)

N-Hydroxymethylphthalimide: Phthalimide, N-(hydroxymethyl)- (8); 1H-Isoindole-1,3-(2H)-dione, 2-(hydroxymethyl)- (9); (118-29-6)

Hydrobromic acid (8,9); (10035-10-6)

Diethyl phthalimidomethylphosphonate: Phosphonic acid, (phthalimidomethyl)-, diethyl ester (8); Phosphonic acid, [(1,3-dihydro-1,3-dioxo-2H-isoindol-2-yl)methyl]-, diethyl ester (9); (33512-26-4)

Triethyl phosphite: Phosphorous acid, triethyl ester (8,9); (122-52-1)

Hydrazine (8,9); (302-01-2)

Diethyl aminomethylphosphonate: Phosphonic acid, (aminomethyl)-, diethyl ester (9); (50917-72-1)

Benzaldehyde (8,9); (100-52-7)

Butyllithium: Lithium, butyl- (8,9); (109-72-8)

Acetophenone (8); Ethanone, 1-phenyl- (9); (98-86-2)

2-Phenyl-N-(phenylmethylene)-1-propen-1-amine: 1-Propen-1-amine, 2-phenyl-N-(phenylmethylene)- (9); (64244-34-4)

Allyl bromide: 1-Propene, 3-bromo- (8,9); (106-95-6)

Phosphorous tribromide: Phosphorus bromide (8); Phosphorous tribromide (9); (7789-60-8)

Diethyl isocyanomethylphosphonate: Phosphonic acid, (isocyanomethyl)-, diethyl ester (9); (41003-94-5)

Dimethyl diazomethylphosphonate: Phosphonic acid, (diazomethyl)-, dimethyl ester (8,9); (27491-70-9)

4-ACETOXYAZETIDIN-2-ONE: SYNTHESIS OF A KEY BETA-LACTAM INTERMEDIATE BY A [2 + 2] CYCLOADDITION ROUTE

(2-Azetidinone, 4-(acetyloxy)-)

H₂C=CH–OAc + O=C=NSO₂Cl → [β-lactam-OAc with N-SO₂Cl] → β-lactam-OAc with NH

Submitted by Stuart J. Mickel,[1] and modified by Chi-Nung Hsiao and Marvin J. Miller.[2]
Checked by Daniel A. Aguilar, John Czepiel, and Gabriel Saucy.

1. Procedure

A 500-mL, four-necked, round-bottomed flask equipped with a mechanical stirrer, rubber septum with a nitrogen source, thermometer, and a pressure-equalized dropping funnel is charged with 150 mL (140 g, 1.63 mol) of vinyl acetate (Notes 1 and 2). Stirring is initiated and the flask content is cooled in an ice-water bath to 3°C. Chlorosulfonyl isocyanate (25 mL, 40 g, 0.28 mol) (Notes 2 and 3) is added as rapidly as possible from the addition funnel while maintaining the temperature at less than 5°C. The cooling bath is removed and the temperature is allowed to rise to 10°C. At this point an exothermic reaction begins. Intermittent cooling is required as the temperature is kept at 10-15°C for 40 min. The dark red mixture is then cooled to -40°C in a dry ice-acetone bath.

A 1.0-L, three-necked flask equipped with a thermometer, mechanical stirrer, and a septum cap is charged with a mixture of 67 g (0.80 mol) of sodium bicarbonate, 71.5 g (0.69 mol) of sodium bisulfite, and 200 mL of water. This mixture is cooled in a dry ice-acetone bath to -20°C with vigorous stirring. Immediately (Note 4) the reaction mixture is added dropwise via cannula at a rate such that the temperature remains at -10°C. This addition takes 30-40 min. When approximately half of the reaction solution has been added, an additional 35.7 g (0.34 mol) of sodium bisulfite is added to the aqueous quench mixture. After the addition is complete, the mixture is stirred for an additional 40 min at -10°C. The light yellow mixture (pH 7) is extracted with three 500-mL portions of chloroform (Note 5). The combined extracts are dried over magnesium sulfate and concentrated on a rotary evaporator at 40°C/70 mm (Note 6). Final solvent removal with a vacuum pump gives a two-phase, oily mixture. The mixture is stirred with three 100-mL portions of hexane, and the hexane extracts are decanted (Note 7) and discarded. Removal of the final traces of solvent with a vacuum pump gives 16.1-22.8 g (44-62% yield based on chlorosulfonyl isocyanate) of a light orange oil which slowly solidifies upon standing at -20°C. The resulting solid melts at 34°C (Note 8).

2. Notes

1. Commercial vinyl acetate (Aldrich Chemical Company, Inc.) was used directly without purification. The checkers observed that distilled vinyl acetate afforded slightly higher yields and improved product purity (Note 7).

2. The volume of reagents used was determined by cannulation into the graduated addition funnel before charging into the reaction flask.

3. Aldrich Chemical Company, Inc. chlorosulfonyl isocyanate was used directly without purification.

4. The mixture tended to freeze if allowed to stand at -20°C.

5. Filtration of the mixture through sintered glass aided in breaking emulsions.

6. Room temperature is even more satisfactory although the concentration takes longer. Heat leads to decomposition of the product.

7. The hexane-soluble impurity is believed to originate in the vinyl acetate. This purification may not be necessary in all cases.

8. The material prepared by this route contains a trace of a yellow impurity (vinyl acetate polymer?). However, the impurity is not detected in the ^{1}H NMR, ^{13}C NMR, or mass spectrum of the product. Very careful column chromatography is required to remove the color and in the hands of the submitters 4-acetoxyazetidin-2-one,[4] prepared by the above method, is adequate for any further manipulation. High vacuum distillation may be employed to obtain a colorless sample (bp 80-82°C/10^{-3} mm); however, extensive losses occur. The spectra are as follows: IR (CHCl$_3$) cm^{-1}: 3350 (NH), 1790 (beta-lactam C=O), 1730 (acetate C=O); ^{1}H NMR (CDCl$_3$) δ: 2.03 (s, 3 H, OCOCH$_3$), 3.00 (d of d, 1 H, J_{3b-3} = 15.0, J_{3b-4} = 1.5), 3.28 (d of d, 1 H, J_{3a-3b} = 15.0, J_{3a-4} = 4.6, H$_{3a}$), 5.81 (d of d, 1 H, J_{4-3a} = 4.6, J_{4-3b} = 1.6), 7.4-7.1 (br, 1 H, NH); ^{13}C NMR (CDCl$_3$) δ (off resonance multiplicity, assignment): 45.0 (t, C3), 73.2 (d, C4), 166.4 (s, C2).

Thin layer chromatographic analysis of 4-acetoxyazetidin-2-one was carried out on E. Merck Silica gel F254 plates by elution with ethyl acetate. The hexane-soluble impurity (R$_f$ 0.67) was detected by shortwave UV. 4-Acetoxyazetidin-2-one (R$_f$ 0.38) was detected by exposure of the plate for 5 min to chlorine gas followed by spraying with TDM solution (Note 9) and heating with a hot air gun.

9. TDM spray solution was prepared as follows: Solution A: 2.5 g of 4,4'-tetramethyldiaminodiphenylmethane (TDM) was dissolved in 10 mL of glacial acetic acid and diluted with 50 mL of H_2O. Solution B: 5 g of potassium iodide was dissolved in 100 mL of H_2O. Solution C: 0.3 g of ninhydrin was dissolved in 90 mL of H_2O diluted with 10 mL of glacial acetic acid. Solutions A and B and 1.5 mL of solution C were mixed and stored in a brown bottle.

3. Discussion

Within the current synthetic effort in beta-lactam chemistry, 4-acetoxyazetidin-2-one and its derivatives play an important role in the total synthesis of many conventional beta-lactams and their analogues.[3-6] There is therefore a requirement for a simple large-scale preparative method for this key intermediate. This synthesis is a modification of that reported by Clauss et al.[7]

1. School of Chemistry, University of Bath, Claverton Down, Bath, BA2 7AY, England. We thank the SERC for support.
2. Department of Chemistry, University of Notre Dame, Notre Dame, IN 46556.
3. Sammes, P. G. "Topics in Antibiotic Chemistry"; Wiley: Chichester, U.K., 1980; Vol. 4.
4. Campbell, M. M.; Carruthers, N. I.; Mickel, S. J.; Winton, P. M. *J. Chem. Soc., Chem. Commun.* **1984**, 200.
5. Bentley, P. H.; Hunt, E. *J. Chem. Soc., Perkin Trans. 1* **1980**, 2222.
6. Reider, P. J.; Grabowski, E. J. *J. Tetrahedron Lett.* **1982**, *23*, 2293.
7. Clauss, K.; Grimm, D.; Prossel, G. *Justus Liebigs Ann. Chem.* **1974**, 539.

Appendix

Chemical Abstracts Nomenclature (Collective Index Number);

(Registry Number)

4-Acetoxyazetidin-2-one: 2-Azetidinone, 4-hydroxy-acetate (ester) (8); 2-Azetidinone, 4-(acetyloxy)- (9); (28562-53-0)

Vinyl acetate: Acetic acid vinyl ester (8); Acetic acid ethenyl ester (9); (108-05-4)

Chlorosulfonyl isocyanate: Sulfuryl chloride isocyanate (9); (1189-71-5)

1,3-DIMETHYL-3-METHOXY-4-PHENYLAZETIDINONE

(2-Azetidinone, 3-methoxy-1,3-dimethyl-4-phenyl-)

A. $Cr(CO)_6 + CH_3Li \longrightarrow (CO)_5Cr=C(OLi)(CH_3) \xrightarrow{(CH_3)_3O^+BF_4^-} (CO)_5Cr=C(OCH_3)(CH_3)$

1

B. $(CO)_5Cr=C(OCH_3)(CH_3) + CH_3N=C(H)(Ph) \xrightarrow[\text{pet. ether}]{h\nu} \xrightarrow[\text{air}]{h\nu}$ [β-lactam product]

2

Submitted by Louis S. Hegedus, Michael A. McGuire, and Lisa M. Schultze.[1]
Checked by Ming Chang P. Yeh and Martin F. Semmelhack.

1. Procedure

A. *[(Methyl)(methoxy)carbene]pentacarbonyl chromium(0).*[2] A 1-L, two-necked, round-bottomed flask equipped with magnetic stirring bar, 100-mL addition funnel, reflux condenser, and gas inlet is charged with 18.7 g (0.085 mol) of chromium hexacarbonyl (Note 1). The apparatus is evacuated (oil pump) and filled with argon (four cycles), and a positive pressure is maintained with an argon-filled balloon on a T-tube. Dry diethyl ether (500 mL) is transferred via cannula into the flask and stirring is commenced. The addition funnel is charged with 60 mL (1.42 M in ether, 0.085 mol) of methyllithium via cannula and rapid dropwise addition is begun. The methyllithium is added over a 15-min period during which time the solution turns from bright yellow to dark brown. The solution is heated at reflux for approximately 1.5 hr. After the solution is cooled, the solvent is removed by rotary evaporation. The dark brown residual solid is taken up in 80 mL of water (in air), and 13.0 g (0.088 mol) of trimethyloxonium tetrafluoroborate (Note 2) is added over a 30-min period with stirring (Note 3). The mixture is extracted several times with 200-mL portions of cold pentane (Note 4). The combined pentane layers are dried over anhydrous magnesium sulfate and filtered through a bed of Celite. The solution is concentrated by rotary evaporation to approximately 60 mL and is cooled to -20°C under argon. After 1 hr the resulting bright yellow crystals (17.6 g, 83%) are collected and dried at 25°C under reduced pressure for 10 min (Note 5).

B. *1,3-Dimethyl-3-methoxy-4-phenylazetidinone*. A 250-mL Pyrex Erlenmeyer flask is charged with 1.25 g (5.0 mmol) of [(methyl)(methoxy)carbene]pentacarbonyl chromium(0)]. The flask is fitted with a rubber septum, evacuated and filled with argon (four cycles). Dry petroleum ether (175 mL) is transferred via cannula into the flask to produce a dark yellow solution. The solution is charged with 0.59 g (5.0 mmol) of N-methylbenzylideneimine (Note 6). The flask is irradiated with six 20-Watt Vitalites (Note 7). The solution turns brown and heterogeneous within an hour. After 3 days (Note 8) the solution is filtered through a bed of Celite, the precipitate is washed with dry petroleum ether, and the now lighter yellow solution is sealed in a flask, degassed, and irradiated as before. After 5 days of further irradiation, the mixture is filtered and the filtrate is exposed to air and irradiated again until a colorless solution is obtained (approx. 1 day). Filtration through a bed of Celite and removal of solvent by rotary evaporation affords colorless crystals of essentially pure β-lactam **2**. Recrystallization from hexane gives 0.67-0.76 g (65-74% yield), mp 76-77°C (Note 9).

2. Notes

1. Chromium hexacarbonyl was obtained by the checkers from Pressure Chemical Company, Pittsburgh, PA, and used without purification. It can be weighed in air as it is relatively non-volatile and air-stable. The usual precautions appropriate for a potentially toxic metal carbonyl should be employed, but the low volatility makes handling relatively easy.

2. The checkers obtained trimethyloxonium tetrafluoroborate from Alfa Products, Morton/Thiokol, Inc..

3. Meerwein's reagent was added until the pH of the solution was slightly acidic.

4. The carbene complex is slightly air sensitive in solution. The pentane was cooled to 0°C and nitrogen was bubbled through the solvent before use.

5. Longer drying resulted in loss of carbene complex by sublimation. The carbene complex was stored under argon at -20°C. The pure product shows ^{1}H NMR (CDCl$_3$) δ: 4.60 (s, 3 H, OC$\underline{H}_3$) and 2.90 (s, 3 H, C$\underline{H}_3$).

6. The checkers obtained N-methylbenzylideneimine from Aldrich Chemical Company, Inc., and used it without purification. It was added as a neat liquid, via syringe.

7. Vitalites were obtained by the checkers from a local hardware store. They were arranged horizontally, in banks of two in a way to provide maximum illumination of the flask. Aluminum foil was used generously around the outside of the lights in order to minimize light loss.

8. Precipitate forms and reduces light intensity in the solution. The complete conversion of reactants can be accelerated by more frequent filtration and by using sunlight in place of the Vitalites. The submitters

were successful using the Vitalites with five filtrations over a 72-hr period. The checkers found that the reaction was incomplete under these conditions and the β-lactam must be purified by chromatography (silica gel column, elution with 1:1 ethyl acetate:hexane) in order to remove residual benzaldehyde and other minor impurities.

9. The product has the following spectral properties: ^{1}H NMR (CDCl$_3$) δ: 1.60 (s, 3 H, CH$_3$); 2.80 (s, 3 H, NCH$_3$); 3.03 (s, 3 H, OCH$_3$); 4.35 (s, 1 H, CH); 7.28 (s, 5 H, ArH); IR (CHCl$_3$) cm^{-1}: 1750.

3. Discussion

The procedure described is an efficient conversion of imines to β-lactams.[3] It is very general, and imines such as thiazolines, benzothiazines, dihydroisoquinoline, and quinoline itself, as well as simple aldehyde and ketone imines are converted to β-lactams in fair to good yield. The reaction is stereospecific, producing only one diastereoisomer of the β-lactam. The chromium carbene complex is easy to prepare on a large scale, to store, and to handle, since it is air stable as a solid. The β-lactam forming-reaction proceeds under very mild conditions and requires only the most simple glassware and either sunlight or commercially available fluorescent tubes which duplicate the spectrum of sunlight, (e.g., Vitalite). Product isolation consists of simple filtration and solvent removal. The procedure produces β-lactams containing heteroatom substituents at the 3-position. It is complementary or superior to existing methods for the conversion of imines to β-lactams involving ketenes,[4] acid chlorides and base,[5] or ketene silyl acetals.[6]

1. Department of Chemistry, Colorado State University, Fort Collins, CO 80523.
2. This is essentially the procedure developed by Fischer: Aumann, R.; Fischer, E. O. *Chem. Ber.* **1968**, *101*, 954.
3. McGuire, M. A.; Hegedus, L. S. *J. Am. Chem. Soc.* **1982**, *104*, 5538; Hegedus, L. S.; McGuire, M. A.; Yijun, C.; Schultze, L. M., submitted for publication in *J. Am. Chem. Soc.*
4. (a) Staudinger, H. *Justus Liebigs Anal. Chem.* **1907**, *356*, 51; (b) Holley, A. D.; Holley, R. W. *J. Am. Chem. Soc.* **1951**, *73*, 3172; (c) Moore, H. W.; Hughes, G. *Tetrahedron Lett.* **1982**, *23*, 4003.
5. (a) Sheehan, J. C.; Ryan, J. J. *J. Am. Chem. Soc.* **1951**, *73*, 1204, 4367; (b) Sheehan, J. C.; Buhle, E. L.; Corey, E. J.; Laubach, G. D.; Ryan, J. J. *J. Am. Chem. Soc.* **1950**, *72*, 3828; (c) Shridhar, D. R.; Ram, B.; Narayana, V. L. *Synthesis* **1982**, *63*; (d) Miyake, M.; Kirisawa, M.; Tokutake, N. *Synthesis*, **1982**, 1053; (e) For a review see: Mukerjee, A. K.; Srivastava, R. C. *Synthesis* **1973**, 327.
6. (a) Ojima, I.; Inaba, S.-i.; Yoshida, K. *Tetrahedron Lett.* **1977**, 3643; (b) Ojima, I.; Inaba, S.-i. *Tetrahedron Lett.* **1980**, *21*, 2077.

Appendix

Chemical Abstracts Nomenclature (Collective Index Number);

(Registry Number)

1,3-Dimethyl-3-methoxy-4-phenylazetidinone: 2-Azetidinone, 3-methoxy-1,3-dimethyl-4-phenyl- (11); (82918-98-7)

[(Methyl)(methoxy)carbene]pentacarbonyl chromium(0): Chromium, pentacarbonyl(1-methoxyethylidene)-, (8); Chromium, pentacarbonyl(1-methoxyethylidene)-, (OC-6-21)-, (9); (20540-69-6)

Chromium hexacarbonyl: Chromium carbonyl (8); Chromium carbonyl (OC-6-11)- (9); (13007-92-6)

Methyllithium: Lithium, methyl- (8,9); (917-54-4)

Trimethyloxonium tetrafluoroborate: Oxonium, trimethyl-, tetrafluoroborate (1-) (8,9); (420-37-1)

N-Methylbenzylideneimine: Methylamine, N-benzylidene- (8); Methanamine, N-(phenylmethylene)- (9); (622-29-7)

4-NITROINDOLE

(Indole, 4-nitro-)

A. [2-methyl-3-nitroaniline] + HC(OEt)₃, TsOH → [ethyl N-(2-methyl-3-nitrophenyl)formimidate]

B. [ethyl N-(2-methyl-3-nitrophenyl)formimidate] + $(CO_2Et)_2$, KOEt, DMF, DMSO → 4-nitroindole

Submitted by Jan Bergman and Peter Sand.[1]
Checked by Cynthia A. Smith and Andrew S. Kende.

1. Procedure

A. Ethyl N-(2-methyl-3-nitrophenyl)formimidate. A 1-L, one-necked, round-bottomed flask, fitted with a Claisen condenser protected from moisture with a drying tube, is charged with 200 g (1.35 mol) of freshly distilled triethyl orthoformate, 1 g of p-toluenesulfonic acid and 152 g (1 mol) of 2-methyl-3-nitroaniline (Notes 1 and 2). The solution is heated to 120°C and all of the ethanol formed is continuously distilled off during ca. 1 hr. Fractional vacuum distillation of the residue gives at 156-158°C/6 mm, the imidate ester, 184 g (88%), as a light yellow, solidifying oil, mp 57-58°C.

B. 4-Nitroindole. To a solution of 22 g (0.15 mol) of diethyl oxalate in 50 mL of dry dimethylformamide in a 200-mL beaker is added, under cooling, 11 g (0.13 mol) of potassium ethoxide with vigorous stirring (Notes 3 and 4). The solution is immediately (within a few seconds) poured into a 250-mL flask containing a solution of 20.8 g (0.10 mol) of ethyl N-(2-methyl-3-nitrophenyl)formimidate in 75 mL of dry dimethyl sulfoxide (Note 5). The resulting deep red solution is stirred for 1 hr at ca. 40°C (Notes 6 and 7). The solution is then transferred into a 1-L beaker and water is added under stirring at a rate which gives smooth precipitation of 4-nitroindole. The product is filtered off and dried giving 16.3 g (ca. 100%) of a brownish-yellow solid, mp 195-201°C (subl.), which is sublimed at 170°C/0.5 mm giving 11.5 g (71%) of yellow crystals, mp 204-205°C (subl.) (Note 8).

2. Notes

1. 2-Methyl-3-nitroaniline and triethyl orthoformate were purchased from Fluka AG.
2. Trimethyl orthoformate is not suitable for this preparation because of sideproduct formation.
3. Diethyl oxalate was purchased from Merck and Company, Inc., and was used without further purification. Potassium ethoxide was purchased from Alfa Products, Morton/Thiokol Inc. or preferably was prepared from potassium metal and absolute ethanol.
4. The diethyl oxalate/potassium ethoxide complex can also be prepared by adding the oxalic ester to an ethanolic solution of potassium ethoxide and evaporating the solvent. However, this complex is less active and is difficult to store.

5. Dimethyl sulfoxide (DMSO) prevents precipitation of intermediate salts, which can also be achieved by using a larger volume of dimethylformamide (DMF) (ca. 200 mL). Attempts to prepare the diethyl oxalate/potassium ethoxide complex in DMSO have not been successful (i.e., it is not active).

6. At elevated temperatures (e.g., above 40°C) by-products are formed.

7. The reaction can be monitored by TLC (CH_2Cl_2). The spots were developed with an ethanolic solution of p-dimethylaminobenzaldehyde/HCl. The product gave a bright red spot at R_f 0.5 and the imidate ester gave a yellow spot at R_f 0.6. Addition of small portions of diethyl oxalate/potassium ethoxide complex was continued if the starting material was not consumed after the initial reaction period.

8. Crude 4-nitroindole can also be purified by recrystallization from methanol, ethanol, or acetonitrile giving brownish-yellow crystals, mp 204-206°C.

3. Discussion

This procedure illustrates the synthesis of 4-nitroindoles; the present method can easily be extended to the 2-alkyl derivatives (using other ortho esters), 5-, 6- and/or 7-substituted derivatives and 1-alkyl derivatives (from the corresponding N-alkylanilides).[2,3] Other published preparations of 4-nitroindole (e.g., ref. 4) are of no practical value.

The mechanism of the formation of 4-nitroindole parallels the Reissert indole synthesis[5] and is discussed in references 2 and 3.

1. Department of Organic Chemistry, Royal Institute of Technology, S-100 44 Stockholm, Sweden.
2. Bergman, J.; Sand, P.; Tilstam, U. *Tetrahedron Lett.* **1983**, *24*, 3665.
3. Bergman, J.; Sand, P., to be published.
4. Somei, M.; Inoue, S.; Tokutake, S.; Yamada, F.; Kaneko, C. *Chem. Pharm. Bull.* **1981**, *29*, 726.
5. Reissert, A. *Chem. Ber.* **1897**, *30*, 1030.

Appendix
Chemical Abstracts Nomenclature (Collective Index Number); (Registry Number)

4-Nitroindole: Indole, 4-nitro- (9); (4769-97-5)

Triethyl orthoformate: Orthoformic acid, triethyl ester (8); Ethane, 1,1',1"-[methylidynetris(oxy)]tris- (9)- (122-51-0)

p-Toluenesulfonic acid monohydrate (8); Benzenesulfonic acid, 4-methyl-, monohydrate (9); (6192-52-5)

2-Methyl-3-nitroaniline: o-Toluidine, 3-nitro- (8); Benzeneamine, 2-methyl-3-nitro- (9); (603-83-8)

Diethyl oxalate: Oxalic acid, diethyl ester (8); Ethanedioic acid, diethyl ester (9); (95-92-1)

Dimethyl sulfoxide: Methyl sulfoxide (8); Methane, sulfinylbis- (9); (67-68-5).

SYNTHESIS OF MACROCYCLIC SULFIDES USING CESIUM THIOLATES:

1,4,8,11-TETRATHIACYCLOTETRADECANE

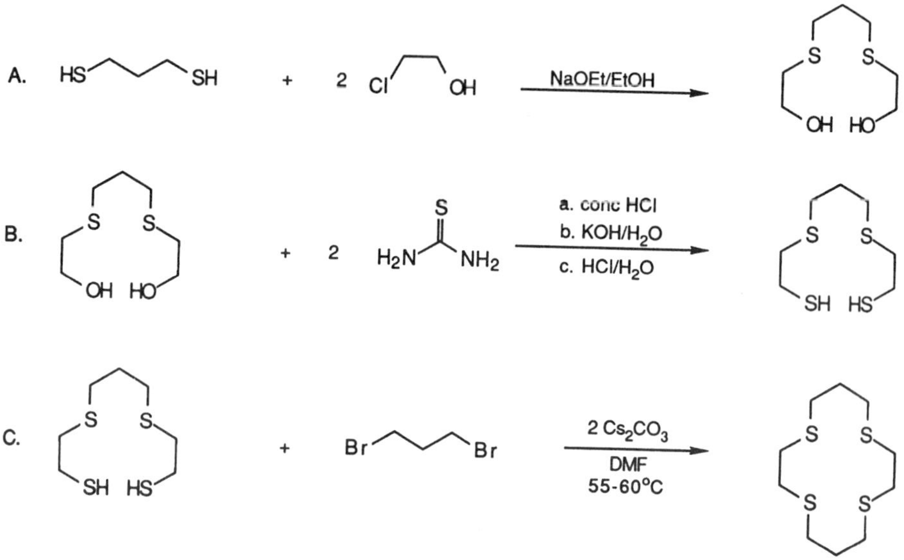

A. HS~~~SH + 2 Cl~~OH →(NaOEt/EtOH) [macrocycle with OH HO]

B. [macrocycle OH HO] + 2 H$_2$N-C(S)-NH$_2$ →(a. conc HCl; b. KOH/H$_2$O; c. HCl/H$_2$O) [macrocycle SH HS]

C. [macrocycle SH HS] + Br~~~Br →(2 Cs$_2$CO$_3$, DMF, 55-60°C) [tetrathia macrocycle]

Submitted by J. Buter and Richard M. Kellogg.[1]
Checked by Joseph M. Salvino and Bruce E. Smart.

1. Procedure

A. *3,7-Dithianonane-1,9-diol.*[2] A 500-mL, three-necked, round-bottomed flask is fitted with a mechanical stirrer, a reflux condenser attached to a nitrogen inlet, and a pressure-equalizing dropping funnel. The flask is flushed with nitrogen and charged with 250 mL of absolute ethanol. The ethanol is stirred and 5.75 g (0.25 mol) of sodium metal is cautiously added. After the sodium dissolves, the solution is warmed to 45-50°C and 13.5 g (0.125 mol) of 1,3-propanedithiol (Note 1) is added dropwise over a period of 15 min. To the resulting solution is added dropwise 20.1 g (0.25 mol) of

2-chloroethanol (Note 1) and the mixture is refluxed for 3-4 hr. The mixture is then allowed to cool to room temperature and is filtered. The filtrate is concentrated on a rotary evaporator to a viscous liquid which is distilled to give 17.3-20.0 g (71-82%) of 3,7-dithianonane-1,9-diol, bp 200°C (1.5 mm) (Note 2).

B. *3,7-Dithianonane-1,9-dithiol.*[2] In a 1-L, round-bottomed flask equipped with a reflux condenser and a magnetic stirring bar are placed 35.0 g (0.178 mol) of 3,7-dithianonane-1,9-diol, 30.0 g (0.394 mol) of thiourea (Note 3), and 94 mL of concd hydrochloric acid. The mixture is stirred and refluxed for 12 hr. The resulting solution is cooled in an ice bath and 67 g (1.2 mol) of potassium hydroxide dissolved in 400 mL of water is added cautiously. The mixture is then refluxed for 3 hr. The resulting two-phase system is cooled to room temperature and the upper aqueous phase is decanted from the oily organic layer. The aqueous phase is acidified with dilute hydrochloric acid and extracted with 300 mL of ether. The ethereal extract is combined with the organic layer from the reaction mixture and this solution is dried over anhydrous magnesium sulfate. The drying agent is removed by filtration and the filtrate is concentrated on a rotary evaporator. The residual liquid is distilled to give 21.4 g (53%) of 3,7-dithianonane-1,9-dithiol, bp 159-162°C (1.2 mm) (Note 4).

C. *1,4,8,11-Tetrathiacyclotetradecane.* A dry, 3-L, three-necked, round-bottomed flask is equipped with a double-necked adapter for a thermometer and the 250-mL addition funnel shown in Figure 1 (Note 5), a reflux condenser, and a mechanical stirrer with a 7-cm blade of Teflon. The entire system is kept under positive nitrogen pressure. The flask is charged with 2.2 L of N,N-dimethylformamide and 13.04 g (40 mmol) of cesium carbonate (Note 6). The mixture is stirred and heated to 55-60°C.

A solution of 9.12 g (40 mmol) of 3,7-dithianonane-1,9-dithiol and 8.08 g (40 mmol) of 1,3-dibromopropane (Note 6) in 300 mL of N,N-dimethylformamide is prepared. Half of this solution is placed in the addition funnel and added to the well-stirred suspension of cesium carbonate in N,N-dimethylformamide over a period of 6-9 hr. The reaction mixture is then charged with another 13.04 g (40 mmol) of cesium carbonate and the second half of the solution of 3,7-dithianonane-1,9-dithiol and 1,3-dibromopropane is added over a period of 6-9 hr (Note 7). After the addition is complete, the reaction mixture is allowed to cool to room temperature. The N,N-dimethylformamide is distilled off as completely as possible under reduced pressure (Note 8). The residue is taken up in 300 mL of dichloromethane and washed once with 200 mL of a saturated solution of sodium chloride. The organic layer is dried over anhydrous magnesium sulfate. The drying agent is removed by filtration and the filtrate is concentrated on a rotary evaporator to a light yellow crystalline mass. This is taken up in 200 mL of boiling 96% ethanol and the hot liquid is decanted. The remaining sediment is boiled with 125 mL of 96% ethanol and again decanted. The two ethanol solutions are combined and stored at 10°C overnight. The white crystalline product that separates is isolated by filtration and dried. There is obtained 6.20-6.60 g (58-62%) of 1,4,8,11-tetrathiacyclotetradecane, mp 118-119°C [lit.[2] mp 119-120°C] (Note 9).

2. Notes

1. 1,3-Propanedithiol and 2-chloroethanol were obtained from the Aldrich Chemical Company, Inc.

2. The submitters report obtaining 19.6-24.5 g (80-100%) of product, bp 179-181°C (0.5 mm). The product obtained by the checkers is pure by NMR and shows the following spectrum: ^{1}H NMR (CDCl$_3$) δ: 1.86 (quintet, 2 H, J = 7, CC$\underline{H}_2$), 2.17 (br s, 2 H, OH), 2.65 (t, 4 H, J = 7, SC$\underline{H}_2$), 2.73 (t, 4 H, J = 6, SC$\underline{H}_2$), 3.73 (t, 4 H, J = 6, OC$\underline{H}_2$).

3. The checkers purchased thiourea from the Aldrich Chemical Company, Inc.

4. The submitters report bp 159-161°C (0.5 mm) for the product and yields of 50-70% for reactions run with 0.1 mol of 3,7-dithianonane-1,9-diol and 0.2 mol of thiourea. The product obtained by the checkers is pure by NMR and shows the following spectrum: ^{1}H NMR (CDCl$_3$) δ: 1.63-2.07 (m, 4 H, CC$\underline{H}_2$, S$\underline{H}$), 2.50-2.83 (m, 12 H, SC$\underline{H}_2$).

5. The device illustrated in Figure 1 allows ready adjustment of addition rates without significant clogging. A is a conical ground glass receiver for the ground glass tapered end of a 7-mm diameter glass rod. The rod is turned in the tapered receiver to attain the desired rate of addition. A ratchet device attached to the top of the addition funnel holds the rod in place and measures its rotation. Outlet B is connected to a mercury bubbler and nitrogen is introduced via C.

6. The checkers obtained N,N-dimethylformamide, cesium carbonate, and 1,3-dibromopropane from the Aldrich Chemical Company, Inc. The N,N-dimethylformamide was distilled and stored over molecular sieves (4 Å) prior to use.

7. The addition rate is sufficiently slow that virtually no starting material remains. The procedure avoids the need for excessively large volumes of solvent that are normally required in high dilution reactions.

8. A Büchi rotary evaporator attached to a vacuum pump is used. A pressure of 1-2 mm is maintained and the flask is heated in a water bath to a maximum temperature of 60°C.

9. The submitters report obtaining 6.97-8.00 g (65-70%) of product, mp 118-119.5°C. The product shows the following spectral properties: IR (KBr) cm^{-1}: 2930, 1430, 1338, 1270, 1205, 1138, 692; ^{1}H NMR (CDCl$_3$) δ: 1.90 (quintet, 4 H, J - 7, CC$\underline{H}_2$), 2.65 (t, 8 H, J = 7, SC$\underline{H}_2$CH$_2$), 2.77 (s, 8 H, SC$\underline{H}_2$).

3. Discussion

The cyclic sulfide 1,4,8,11-tetrathiacyclotetradecane has been prepared in 7.5% yield by reaction of the bis-Na$^+$ salt of 3,7-dithianonane-1,9-dithiol in boiling ethanol with 1,3-dibromopropane.[2] A marked improvement in yield by the use of cesium salts in N,N-dimethylformamide has been documented for other macrocyclic sulfides,[3] as well as macrocyclic lactones,[4,5] and amines.[6] Moreover, the nucleophilic properties of cesium salts have been used to advantage in substitution reactions.[7]

Macrocyclic sulfides, including 1,4,8,11-tetrathiacyclotetradecane, are of interest, especially as ligands for transition metal ions in a variety of different applications.[8-29] The methodology described here provides an efficient entry to many such macrocycles, including chiral ones that act as ligands in transition metal-catalyzed coupling reactions.[29]

1. Department of Organic Chemistry, University of Groningen, Nijenborgh 16, 9747 AG Groningen, The Netherlands.
2. Rosen, W.; Busch, D. H. *J. Am. Chem. Soc.* **1969**, *91*, 4694.
3. Buter, J.; Kellogg, R. M. *J. Org. Chem.* **1981**, *46*, 4481.

4. Kruizinga, W. H.; Kellogg, R. M. *J. Am. Chem. Soc.* **1981**, *103*, 5183.
5. Jouin, P.; Troostwijk, C. B.; Kellogg, R. M. *J. Am. Chem. Soc.* **1981**, *103*, 2091.
6. Vriesema, B. K.; Buter, J.; Kellogg, R. M. *J. Org. Chem.* **1984**, *49*, 110.
7. Kruizinga, W. H.; Strijtveen, B.; Kellogg, R. M. *J. Org. Chem.* **1981**, *46*, 4321.
8. Dann, J. R.; Chiesa, P. P.; Gates, J. W., Jr. *J. Org. Chem.* **1961**, *26*, 1991.
9. Eglinton, G.; Lardy, I. A.; Raphael, R. A.; Sim, G. A. *J. Chem. Soc.* **1964**, 1154.
10. Bradshaw, J. S.; Hui, J. Y.; Haymore, B. L.; Christensen, J. J.; Izatt, R. M. *J. Heterocycl. Chem.* **1973**, *10*, 1.
11. Bradshaw, J. S.; Hui, J. Y.; Chan, Y.; Haymore, B. L.; Izatt, R. M.; Christensen, J. J. *J. Heterocycl. Chem.* **1974**, *11*, 45.
12. Dietrich, B.; Lehn, J. M.; Sauvage, J. P. *J. Chem. Soc., Chem. Commun.* **1970**, 1055.
13. Alberts, A. H.; Annunziata, R.; Lehn, J.-M. *J. Am. Chem. Soc.* **1977**, *99*, 8502.
14. Kahn, O.; Morgenstern-Badaru, I.; Audiere, J. P.; Lehn, J.-M.; Sullivan, S. A. *J. Am. Chem. Soc.* **1980**, *102*, 5935.
15. Frensdorff, H. K. *J. Am. Chem. Soc.* **1971**, *93*, 600.
16. Jones, T. E.; Zimmer, L. L.; Diaddario, L. L; Rorabacher, D. B.; Ochrymowycz, L. A. *J. Am. Chem. Soc.* **1975**, *97*, 7163.
17. Jones, T. E.; Rorabacher, D. B.; Ochrymowycz, L. A. *J. Am. Chem. Soc.* **1975**, *97*, 7485.
18. Dockal, E. R.; Jones, T. E.; Sockol, W. F.; Engener, R. J.; Rorabacher, D. B.; Ochrymowycz, L. A. *J. Am. Chem. Soc.* **1976**, *98*, 4324.

19. Ferris, N. S.; Woodruff, W. H.; Rorabacher, D. B.; Jones, T. E.; Ochrymowycz, L. A. *J. Am. Chem. Soc.* **1978**, *100*, 5939.
20. Black, D. St. C.; McLean, I. A. *Aust. J. Chem.* **1971**, *24*, 1401.
21. Rosen, W.; Busch, D. H. *J. Chem. Soc., Chem. Commun.* **1969**, 148.
22. Travis, K.; Busch, D. H. *J. Chem. Soc., Chem. Commun.* **1970**, 1041.
23. Rosen, W.; Busch, D. H. *Inorg. Chem.* **1970**, *9*, 262.
24. Weber, F.; Vögtle, F. *Justus Liebigs Ann. Chem.* **1976**, 891.
25. Musker, W. K.; Wolford, T. L.; Roush, P. B. *J. Am. Chem. Soc.* **1978**, *100*, 6416.
26. Asmus, K.-D.; Bahnemann, D.; Fischer, Ch.-H.; Veltwisch, D. *J. Am. Chem. Soc.* **1979**, *101*, 5322.
27. Doi, J. T.; Musker, W. K. *J. Am. Chem. Soc.* **1981**, *103*, 1159.
28. Hintsa, E. J.; Hartman, J. R.; Cooper, S. R. *J. Am. Chem. Soc.* **1983**, *105*, 3738.
29. Lemaire, M.; Buter, J.; Vriesema, B. K.; Kellogg, R. M. *J. Chem. Soc., Chem. Commun.* **1984**, 309.

Figure 1

Addition Funnel

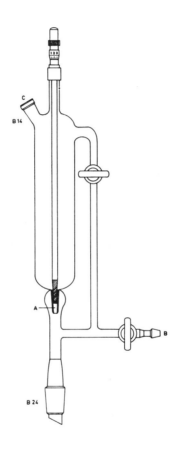

Appendix

Chemical Abstracts Nomenclature (Collective Index Number);

(Registry Number)

1,4,8,11-Tetrathiacyclotetradecane (8,9); (24194-61-4)

1,3-Propanedithiol (8,9); (109-80-8)

2-Chloroethanol (8); Ethanol, 2-chloro- (9); (107-07-3)

3,7-Dithianonane-1,9-diol: Ethanol, 2,2'-(trimethylenedithiol)di- (8,9); Ethanol, 2,2'-[1,3-propanediylbis(thio)]bis- (10); (16260-48-3)

3,7-Dithianonane-1,9-dithiol: Ethanethiol, 2,2'-(trimethylenedithio)di- (8); Ethanethiol, 2,2'-[1,3-propanediylbis(thio)]bis- (9); (25676-62-4)

Thiourea: Urea, thio- (8); Thiourea (9); (62-56-6)

N,N-Dimethylformamide: Formamide, N,N-dimethyl- (8,9); (68-12-2)

Cesium carbonate: Carbonic acid, dicesium salt (8,9); (534-17-8)

1,3-Dibromopropane: Propane, 1,3-dibromo- (8,9); (109-64-8)

ALLYLCARBAMATES BY THE AZA-ENE REACTION:

METHYL N-(2-METHYL-2-BUTENYL)CARBAMATE

(Carbamic acid, (2-methyl-2-butenyl)-, methyl ester)

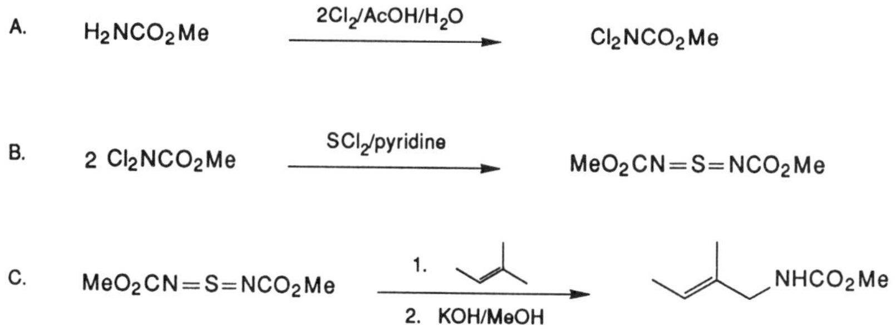

A. H_2NCO_2Me →[$2Cl_2$/AcOH/H_2O] Cl_2NCO_2Me

B. $2\ Cl_2NCO_2Me$ →[SCl_2/pyridine] $MeO_2CN=S=NCO_2Me$

C. $MeO_2CN=S=NCO_2Me$ →[1. (2-methyl-2-butene) 2. KOH/MeOH] (2-methyl-2-butenyl)$NHCO_2Me$

Submitted by Günter Kresze, Hans Braxmeier, and Heribert Münsterer.[1]
Checked by N. Laxma Reddy and Ian Fleming.

1. Procedure

Caution! All three parts of this preparation should be performed in a well-ventilated hood. The reagents and the products of parts A and B are toxic and unpleasant substances.

A. *Methyl N,N-dichlorocarbamate.* A 2-L, two-necked, round-bottomed flask equipped with a magnetic stirrer, gas inlet and gas outlet is charged with 84 g (1.1 mol) of methyl carbamate, 210 g (2.6 mol) of sodium acetate, 21 g (0.35 mol) of glacial acetic acid and 400 mL of water and cooled to -10° to -15°C (Note 1). About 175 g (2.5 mol) of chlorine is condensed in a calibrated Schlenk tube (Note 2) cooled with dry ice/methanol. The cooling bath is replaced by an ice-water bath (Note 3) and chlorine is passed slowly

(Note 4) (over 2 hr, Note 5) at a constant rate into the solution, which is vigorously stirred with a magnetic stirrer. The mixture is transferred to a separatory funnel and the yellow oil which settles is run off. The yellow oil is washed with three subsequent 50-mL portions of a 20% aqueous sodium chloride solution and dried over anhydrous magnesium sulfate. The crude product is then transferred to a distillation apparatus where it is kept under reduced pressure (11-15 mm) at room temperature for about 20 min. The bath temperature is slowly raised to a maximum of 60°C. The product distils at 43°C (11 mm) to give 102-118 g (63-73%, based on methyl carbamate) of a heavy yellow oil (Note 6, Note 7, Note 8).

B. N^1,N^2-Bis(methoxycarbonyl)sulfur diimide. A 250-mL, three-necked, round-bottomed flask, equipped with a gas outlet stop-cock, a thermometer, and a pressure-equalizing dropping funnel and containing a magnetic stirrer bar is purged with dry nitrogen and charged with 53.0 g (0.37 mol) of methyl N,N-dichlorocarbamate and 0.2 mL of pyridine. The dropping funnel is filled with 20.6 g (0.2 mol) of freshly distilled sulfur dichloride (Note 9). The whole apparatus is closed and connected by way of the gas outlet to a paraffin oil-filled valve (Note 10). About 2 mL of sulfur dichloride is then added dropwise into the stirred flask. When the evolution of chlorine has started (usually after about 5 min) (Note 11), the remainder of the sulfur dichloride is added at such a rate that 3-5 bubbles of chlorine per second are evolved (Note 12), without allowing the temperature of the reaction mixture to exceed 35°C for more than short intervals. The addition takes 1.5 to 2 hr. After the addition is completed and the evolution of gas has slowed down significantly, volatile materials are removed by stirring the mixture at 60°C under reduced pressure (11-15 mm) for about 10 min (Note 13). Further removal is accomplished at room temperature at 0.01-0.05 mm for approximately 1 hr.

The product (33.0-35.0 g), a moisture-sensitive, viscous, yellow oil, is used directly without further purification (Note 14).

C. *Methyl N-(2-methyl-2-butenyl)carbamate.* A 250-mL, two-necked flask, equipped with a gas-outlet stop-cock and a pressure-equalizing dropping funnel, and containing a magnetic stirrer bar, is flushed with dry nitrogen and charged with a solution of N^1,N^2-bis(methoxycarbonyl)sulfur diimide from Part B in 30 mL of dry chloroform. The solution is cooled with ice water, and 14.1 g (0.20 mol) of 2-methyl-2-butene is slowly dropped in with stirring over 1 hr. After the completion of the addition, the reaction mixture is stirred for another 10 hr at room temperature. The funnel is replaced by a distillation unit and most of the solvent is removed (11-15 mm, bath temperature 30°C). To the residue are added 240 mL of a 10% solution of potassium hydroxide in methanol and about 10 mL of water. The mixture is stirred for 3 hr at room temperature, and any precipitate is filtered off (Note 15). The residue is washed with 50 mL of methanol, and the filtrate, a red solution, is concentrated at 40°C (11-15 mm). The residue is taken up in 450 mL of ether and washed with four 100-mL portions of water. The ethereal layer is dried with anhydrous magnesium sulfate and treated with charcoal until the color has changed from red to yellow, whereupon the solvent is removed under reduced pressure. The resultant oil is distilled at 70-75°C (3 mm), 62-64°C (0.2 mm) to give the colorless product, yield 10.7-13.7 g (43-52%, based on N,N-dichlorocarbamate) (Notes 16, 17).

2. Notes

1. The solution of methyl carbamate, sodium acetate and acetic acid is best prepared at room temperature. Upon cooling small amounts of precipitate form, but this precipitate dissolves again during the reaction.

2. An amount of 175 g of condensed chlorine corresponds to a volume of about 115 mL. A dry ice-acetone condenser is fitted above the Schlenk tube.

3. A safety flask of at least 1 L should be placed between the Schlenk tube and the reaction flask.

4. To prevent chlorine from escaping before reaction has taken place, the gas inlet tube is immersed as deeply as possible into the solution.

5. To follow the rate of evaporation of chlorine, it is helpful to have calibrated the Schlenk tube.

6. The crude product should not be stored, and the distilled product is best kept protected from light at -78°C.

7. During the distillation, the product is best kept cooled in an ice-water bath.

8. The product is a powerful skin irritant. Protective gloves should be worn during the separation, and when transferring the liquid to the distillation apparatus.

9. Sulfur dichloride decomposes at its boiling point. It is best distilled at low pressure, condensing it in a cooled vessel.

10. The valve serves as a bubble counter for monitoring the evolution of gas.

11. It sometimes happens that no significant evolution of chlorine occurs. In these cases the mixture is heated to 30-40°C with a water bath, which is removed after the reaction has started. The checkers found no delay in the evolution of chlorine.

12. This evolution should not be interrupted.

13. An aspirator and a water bath may be used for this purpose. A tube filled with anhydrous calcium chloride should be placed between the aspirator and the flask.

14. The product is best stored protected from light at -78°C. Even under these conditions, it is advisable to use it up within at least 1 month.

15. Sometimes no significant amount of precipitate is formed, in which case filtration may be omitted.

16. During distillation the receiving flask should be cooled with ice water in order to minimize losses. The product is a semisolid material at room temperature and solidifies completely when kept in a refrigerator.

17. The product has the following spectroscopic properties: IR (film) cm^{-1}: 3520 (NH), 1700 (CO) 1530; ^{1}H NMR (250 MHz, d_5-pyridine) δ: 1.50 (d, 3 H, J = 6.7, C$\underline{H}_3$CH), 1.59 (s, 3 H, C$\underline{H}_3$C=), 3.72 (s, 3 H, C$\underline{H}_3$O), 3.90 (d, 2 H, J = 6, C$\underline{H}_2$N), 5.48 (q, 1 H, J = 6.7, $\underline{H}$C=), 8.00 (m, 1 H, $\underline{H}$N), with long-range splitting evident in the fine structure. It appears to be a single stereoisomer (> 97:3).

3. Discussion

The preparation of methyl N,N-dichlorocarbamate is based on work of Toglia and Swern,[2] and the preparation of the sulfurdiimide is based on work of Levchenko.[3] In both cases, we have modified the method and added significant details.

The synthesis of primary allylamines up to 1983 has been reviewed.[4] Our method involves the ene reaction of various aza analogues of sulfur or selenium dioxide (1 + 2 → 3), followed by [2,3]-sigmatropic rearrangement (3 → 4). The use of N-tosyl activating groups, as in our earlier work[5] and that of Sharpless,[6-8] has the disadvantage that the N-sulfonyl group cannot easily be

removed under mild conditions, as Sharpless observed in his synthesis of gabaculine.[8] In the present method, the methoxycarbonyl group is used in place of the sulfonyl group. It is easy to remove, but the diimide (2) is less reactive than the corresponding sulfonyl compound.[9] Nevertheless it reacts at room temperature with a variety of alkenes, such as β-methylstyrene, 2-pentene, and cyclohexene, and more heavily substituted derivatives of these compounds.[10] The urethane group can easily be hydrolyzed or reduced (lithium aluminum hydride) to give the corresponding allylamines.[10]

1. Organisch-Chemisches Institut, Technische Universität München, D-8046 Garching, Germany.
2. Foglia, T. A.; Swern, D. *J. Org. Chem.* **1966**, *31*, 3625.
3. Levchenko, E. S.; Bal'on, Ya. G.; Kirsanov, A. V. *Zh. Org. Khim.* **1967**, *3*, 2068, Engl. transl. 2014; *Chem. Abstr.* **1968**, *68*, 39568.
4. Laurent, A.; Mison, P.; Nafti, A. *Synthesis* **1983**, 685.
5. Schönberger, N.; Kresze, G. *Liebigs Ann. Chem.* **1975**, 1725.
6. Sharpless, K. B.; Hori, T. *J. Org. Chem.* **1976**, *41*, 176.

164

7. Sharpless, K. B.; Hori, T.; Truesdale, L. K.; Dietrich, C. O. *J. Am. Chem. Soc.* **1976**, *98*, 269.
8. Singer, S. P.; Sharpless, K. B. *J. Org. Chem.* **1978**, *43*, 1448.
9. Bussas, R.; Kresze, G. *Liebigs Ann. Chem.* **1980**, 629.
10. Kresze, G.; Münsterer, H. *J. Org. Chem.* **1983**, *48*, 3561.

Appendix
Chemical Abstracts Nomenclature (Collective Index Number); (Registry Number)

Methyl N-(2-methyl-2-butenyl)carbamate: Carbamic acid, (2-methyl-2-butenyl)-, methyl ester (11); (86766-65-6)

Methyl N,N-dichlorocarbamate: Carbamic acid, dichloro-, methyl ester (8,9); (16487-46-0)

Methyl carbamate: Carbamic acid, methyl ester (8,9); (598-55-0)

Chlorine (8,9); (7782-50-5)

N^1,N^2-Bis(methoxycarbonyl)sulfur diimide: Sulfur diimide, dicarboxy-, dimethyl ester (8,9); (16762-82-6)

Sulfur dichloride: Sulfur chloride (8,9): (10545-99-0)

2-Methyl-2-butene: 2-Butene, 2-methyl- (8,9); (513-35-9)

tert-BUTYL-tert-OCTYLAMINE

(2-Pentanamine, N-(1,1-dimethylethyl)-2,4,4-trimethyl-)

A. $(CH_3)_3CCH_2C(CH_3)NH_2$ + $Na_2WO_4 \cdot 2 H_2O/H_2O_2$ ⟶ $(CH_3)_3CCH_2C(CH_3)_2N=O$

B. $(CH_3)_3CCH_2C(CH_3)_2N=O$ + $(CH_3)_3CNHNH_2$ + PbO_2 ⟶ $(CH_3)_3CCH_2C(CH_3)_2\underset{\underset{\displaystyle NC(CH_3)}{|}}{N}-OC(CH_3)$

C. $(CH_3)_3CCH_2C(CH_3)_2\underset{\underset{\displaystyle NC(CH_3)_3}{|}}{N}-OC(CH_3)_3$ + [naphthalene]⁻ Na⁺ ⟶ $(CH_3)_3CCH_2C(CH_3)_2\underset{\underset{\displaystyle NC(CH_3)}{|}}{N}-H$

Submitted by E. J. Corey and Andrew W. Gross.[1]
Checked by Axel H. K. Paul and Clayton H. Heathcock.

1. Procedure

A. *Nitroso-tert-octane*.[2] To a 1-L, three-necked flask equipped with an addition funnel, a mechanical stirrer, and a thermometer are added 120 mL of methanol, 51.7 g of tert-octylamine (0.4 mol) and 90 mL of water containing 1.2 g (0.0028 mol) of the tetrasodium salt of ethylenediaminetetraacetic acid and 2.52 g (0.0076 mol) of sodium tungstate dihydrate. The solution is cooled to 15°C in an ice bath and hydrogen peroxide (361 mL of a 16% solution, 1.7 mol) (Notes 1 and 2) is added over 5 hr. The blue reaction mixture is stirred for an additional 16 hr and the product is extracted with petroleum ether (3 x 50 mL). Unreacted amine is removed by washing twice with 2 N hydrochloric acid. After the blue organic layer is washed with brine, it is dried over $MgSO_4$. Petroleum ether is removed by distillation at atmospheric pressure. Continued distillation of the product affords 29.7 g of nitroso-tert-octane

(52%), bp 45-55°C, 18 mm. Upon standing, the product crystallizes as the colorless dimer, mp 63-65°C. For use in the subsequent reaction the required amount of the dimer is stirred for 1 hr in hexane to establish monomer-dimer equilibrium (Notes 3 and 4).

B. *N-tert-Butyl-N-tert-octyl-O-tert-butylhydroxylamine.* Into a 1-L, three-necked flask equipped with a dropping funnel, mechanical stirrer, and gas inlet tube is placed a solution of 22.8 g (0.16 mol of monomer) of nitroso-tert-octane in 500 mL of hexane. After the mixture is stirred for 1 hr, lead dioxide (132 g, 0.55 mol) (Note 5) is added. To the rapidly stirring mixture is added 48.8 g (0.55 mol) of tert-butylhydrazine (Note 6) dropwise at such a rate as to give brisk but controlled nitrogen evolution (approximately 30 min). Cooling is provided by an ice bath so as to maintain the reaction temperature between 15 and 25°C. Progress of the reaction is monitored visually by the disappearance of the blue nitroso monomer color (Note 7). After the blue color has disappeared (about 1.5 hr), the lead oxides are removed by filtration through a pad of Celite and the residue is washed with ether/hexane (1:1). The filtrate and washes are combined and the solvents are removed at reduced pressure with a rotary evaporator to give a 6:1 mixture of N-tert-butyl-N-tert-octyl-O-tert-butylhydroxylamine and N-tert-octyl-O-tert-butylhydroxylamine (Notes 8 and 9).

C. *tert-Butyl-tert-octylamine.* In a dry 500-mL, three-necked flask equipped with a mechanical stirrer, addition funnel, and nitrogen inlet are placed 25.6 g (0.20 mol) of naphthalene, 250 mL of dry tetrahydrofuran (THF), and 10.8 g (0.47 mol) of sodium pieces. The mixture is stirred at room temperature for 30 min. To the blue-green sodium naphthalenide solution is added the hydroxylamine mixture (Note 10) in 50 mL of THF over 20 min (*Caution: exothermic reaction*). The mixture is stirred for 2.5 hr at room temperature (Note 11). The reaction mixture is carefully decanted from excess sodium and the excess reducing agent is cautiously quenched with isopropyl alcohol. After dilution with 150 mL of hexane, the mixture is acidified with 300 mL of ice-cold 2 N hydrochloric acid, the aqueous layer is separated, and the organic layer extracted twice more with 100-mL portions of 2 N hydrochloric acid. The combined acidic extracts are washed with 80 mL of petroleum ether (Note 12), neutralized with 4 N sodium hydroxide, and extracted with ether (3 x 100 mL). The ether extract is dried over $MgSO_4$ and the solvent is removed with a rotary evaporator. Distillation of the residue gives a small forerun, followed by 18-19 g of tert-butyl-tert-octylamine, bp 79°C (17 mm). The yield is 60-64% based on nitroso-tert-octane (Note 13).

2. Notes

1. Mallinckrodt 30% H_2O_2 was diluted with water to give a solution of 16% H_2O_2 in water.

2. Although it is used in excess, the amount of H_2O_2 used seems to be critical. The checkers found that the use of 2.1 mol of H_2O_2 results in considerable over-oxidation to nitro-tert-octane, resulting in a yield of nitroso-tert-octane of only 40%.

3. Nitroso-tert-octane may also be prepared by oxidation of tert-octylamine with peracetic acid in ethyl acetate, obtained by the submitters from the Union Carbide Corporation.[4] To a 1-L, three-necked flask equipped with a mechanical stirrer and an addition funnel are added 51.7 g of tert-octylamine (0.4 mol), 50 mL of water, and 50 mL of ethyl acetate. The flask is placed in an ice bath and the contents are stirred until the temperature reaches 0-5°C. A solution of peracetic acid in ethyl acetate (3.15 M

solution, 51 mL, 0.16 mol) is added dropwise over a period of 30 min. The blue reaction mixture is stirred at 0°C until the absence of peroxy acid is indicated by starch-iodide test paper. The reaction mixture is transferred to a separatory funnel and diluted with 200 mL of hexane. Unreacted tert-octylamine is removed by washing with 4 N hydrochloric acid. The aqueous washes are backwashed until colorless with 50-mL portions of hexane. The combined, blue organic fractions are dried over sodium sulfate and used directly in the next step. When the nitroso compound is prepared in this manner, isolation is unnecessary. The checkers did not employ this procedure because the peracetic acid/ethyl acetate solution is not commercially available.

4. Care should be taken in distilling the nitroso compound because it is thermally unstable; its half-life is less than 5 min at 150°C.[2]

5. If technical-grade lead dioxide (Fisher Scientific Company) is used a somewhat greater amount must be added to compensate for the decreased Pb(IV) content.

6. tert-Butylhydrazine is conveniently liberated from its hydrochloride (Aldrich Chemical Company, Inc.) by distillation from 40% KOH (bp 104-107°C). The distillate consists of a mixture of tert-butylhydrazine and water. Upon addition of several grams of KOH pellets, the distillate separates into two layers. The upper layer, consisting of slightly wet tert-butylhydrazine is dried over KOH pellets.

7. Additional lead dioxide and/or tert-butylhydrazine may be necessary to complete the reaction.

8. The spectral properties are as follows: N-tert-Butyl-N-tert-octyl-O-tert-butylhydroxylamine ^{1}H NMR (CDCl$_3$) δ: 1.00 (s, 9 H), 1.26 (s, 12 H), 1.30 (s, 3 H), 1.31 (s, 9 H), 1.76 (s, 2 H); N-tert-octyl-O-tert-

butylhydroxylamine ^{1}H NMR (CDCl$_3$) δ: 1.01 (s, 9 H), 1.12 (s, 6 H), 1.15 (s, 9 H), 1.40 (s, 2 H), 4.44 (s, 1 H).

9. The trisubstituted hydroxylamine is sensitive to both acid and heat.

10. The yield of product is greatly reduced if the hydroxylamine mixture is not rigorously freed of solvent from the previous reaction.

11. The submitters report that the sodium naphthalenide color is discharged upon addition of the hydroxylamine mixture, and that completion of the reduction is indicated by reappearance of the characteristic blue-green color of the reagent. The checkers did not observe disappearance of the reagent color.

12. If the acidic solution is not extracted at this point, the final product will be contaminated with 2-3% naphthalene.

13. The ^{1}H NMR spectrum of tert-butyl-tert-octylamine is as follows (CDCl$_3$) δ: 1.02 (s, 9 H, C$\underline{H}_3$), 1.19 (s, 9 H, C$\underline{H}_3$), 1.24 (s, 6 H, C$\underline{H}_3$), 1.44 (s, 2 H, C$\underline{H}_2$).

3. Discussion

In addition to the procedure given here for the oxidation of tert-octylamine to nitroso-tert-octane,[2] the oxidation may be carried out with m-chloroperoxybenzoic acid[3] or with a solution of peroxyacetic acid in ethyl acetate.[4] The lead dioxide oxidation of alkylhydrazines to alkyl radicals appears to have general application. In addition to tert-butylhydrazine, various secondary alkylhydrazines (e.g., bornylhydrazine and menthylhydrazine) have been used to good effect. The reduction of tri-tert-alkylhydroxylamine to the di-tert-alkylamine has also been achieved with sodium in ammonia but the insolubility of the hydrophobic substrate makes this procedure difficult. The use of sodium naphthalenide[5] gives higher yields and is more reproducible.

In addition to the commercially available 2,2,6,6-tetramethylpiperidine,[6] di-tert-alkylamines have been prepared by Rathke[7] by the copper-catalyzed coupling of acetylenic amines with acetylenic chlorides in an improvement of the procedure of Hennion.[8] Di-tert-butylamine has been synthesized by the reaction of 2-methyl-2-nitropropane with sodium, followed by reduction.[9]

The three-step procedure described here illustrates a convenient, general route to di-tert-alkylamines. Extensive purification or isolation of intermediates is not required. The reactions are easily monitored. Only in the final step is the exclusion of air and moisture necessary. It should be noted that tert-butyl-tert-octylamine is considerably more hindered than 2,2,6,6-tetramethylpiperidine. tert-Butyl-tert-octylamine is inert to methyl iodide, while 2,2,6,6-tetramethylpiperidine gives a white precipitate of the pentamethylammonium iodide within minutes upon treatment with methyl iodide at room temperature. The extreme hindrance of this amine has been exploited in the selective deprotonation of carbon acids and in other reactions.[10]

1. Department of Chemistry, Harvard University, Cambridge, MA 02138.
2. Stowell, J. C. *J. Org. Chem.* **1971**, *36*, 3055.
3. Baldwin, J. E.; Qureshi, A. K.; Sklarz, B. *J. Chem. Soc. (C)* **1969**, 1073.
4. Corey, E. J.; Gross, A. W. *Tetrahedron Lett.* **1984**, *25*, 491.
5. Closson, W. D.; Wriede, P.; Bank, S. *J. Am. Chem. Soc.* **1966**, *88*, 1581.
6. Olofson, R. A.; Dougherty, C. M. *J. Am. Chem. Soc.* **1973**, *95*, 581.
7. Kopka, I. E.; Fataftah, Z. A.; Rathke, M. W. *J. Org. Chem.* **1980**, *45*, 4616.
8. Hennion, G. F.; Hanzel, R. S. *J. Am. Chem. Soc.* **1960**, *82*, 4908.
9. Hoffmann, A. K.; Feldman, A. M.; Gelblum, E.; Hodgson, W. G. *J. Am. Chem. Soc.* **1964**, *86*, 639; Back, T. G.; Barton, D. H. R. *J. Chem. Soc., Perkin I* **1977**, 924.

10. Corey, E. J.; Gross, A. W. *Tetrahedron Lett.* **1984**, *25*, 495.

Appendix
Chemical Abstracts Nomenclature (Collective Index Number); (Registry Number)

tert-Butyl-tert-octylamine: 2-Pentanamine, N-(1,1-dimethylethyl)-2,4,4-trimethyl- (11); (90545-94-1)

Nitroso-tert-octane: Pentane, 2,2,4-trimethyl-4-nitroso- (8,9); (31044-98-1)

tert-Octylamine: 2-Pentanamine, 2,4,4-trimethyl- (9); (107-45-9)

Ethylenediaminetetraacetic acid, tetrasodium salt: Glycine, N,N'-1,2-ethanediylbis[N-(carboxymethyl)]-, tetrasodium salt, trihydrate (9); (67401-50-7)

Sodium tungstate dihydrate: Tungstic acid, disodium salt, dihydrate (8,9); (10213-10-2)

Hydrogen peroxide (8,9); (7722-84-1)

N-tert-Butyl-N-tert-octyl-O-tert-butylhydroxylamine (11): 2-Pentanamine, N-(1,1-dimethylethoxy)-N-(1,1-dimethylethyl)-2,4,4-trimethyl- (11); (90545-93-0)

Lead dioxide: Lead oxide (8,9); 1309-60-0)

tert-Butylhydrazine hydrochloride: Hydrazine, tert-butyl, monohydrochloride (8); Hydrazine, (1,1-dimethylethyl)-, monohydrochloride (9); (7400-27-3)

N-tert-Octyl-O-tert-butylhydroxylamine: 2-Pentanamine, N-(1,1-dimethylethoxy)-2,4,4-trimethyl- (10); (68295-32-9)

Naphthalene (8,9); (91-20-3)

Sodium (8,9); (7440-23-5)

(S)-(-)-1-AMINO-2-METHOXYMETHYLPYRROLIDINE (SAMP) AND (R)-(+)-1-AMINO-2-METHOXYMETHYLPYRROLIDINE (RAMP), VERSATILE CHIRAL AUXILIARIES

(1-Pyrrolidinamine, 2-(methoxymethyl)-, (S))
(1-Pyrrolidinamine, 2-(methoxymethyl)-, (R))

Submitted by Dieter Enders, Peter Fey, and Helmut Kipphardt.[1]
Checked by Akira Yanagisawa and Ryoji Noyori.

1. Procedure

A. *(S)-(+)-2-Hydroxymethylpyrrolidine*. In a 4-L, three-necked, round-bottomed flask equipped with a heating mantle (Note 1), an overhead stirrer bearing a two-bladed propeller (ca. 2.5-cm diameter), an effective reflux condenser with a drying tube packed with silica gel, and a plastic stopper are placed 2.5 L of anhydrous tetrahydrofuran (THF, Note 2) and 60 g (1.56 mol) of lithium aluminum hydride ($LiAlH_4$, Note 3). The suspension is heated under reflux for 15 min, the heating mantle is switched off, and 115.1 g (1 mol) of powdered (S)-proline (Note 4) is added in small portions (ca. 2 g, Note 5) to the boiling mixture at such a rate as to maintain reflux. The addition requires ca. 45 min and the contents of the flask are kept boiling for an additional 1 hr. Excess lithium aluminum hydride is then decomposed by cautiously adding a solution of 28 g of potassium hydroxide in 112 mL of water (without external heating) through a pressure-equalizing dropping funnel to the boiling mixture. Upon hydrolysis, white salts precipitate and stirring becomes difficult. After the addition is complete (ca. 25 min), the mixture is refluxed for 15 min and the hot solution is filtered by suction through a large Büchner funnel (18-cm diameter). The precipitate is pressed dry with a beaker. Any remaining prolinol is extracted from the precipitate by refluxing with 1.5 L of tetrahydrofuran for 1 hr under mechanical stirring, followed again by suction filtration. The combined filtrates are concentrated in a 2-L flask at 30°C (Note 6) under reduced pressure to yield 115-125 g of the crude hydroxymethylpyrrolidine as a pale yellow oil.

B. *(S)-(-)-1-Formyl-2-hydroxymethylpyrrolidine*. The 2-L flask containing the crude hydroxymethyl derivative (ca. 1 mol) is equipped with a dropping funnel and a magnetic stirring bar, and cooled to 0°C. Eighty milliliters (1.3 mol) of methyl formate (Note 7) is added over a period of 20 min and stirring is continued for 30 min at 0°C to give a green-colored solution. Excess methyl formate is evaporated at 30°C affording a dark oil, which is taken up in 600 mL of dichloromethane and dried twice by stirring over a sufficient amount of anhydrous sodium sulfate. The drying agent is removed by suction filtration through Celite (Note 8), using a large column, and the filtrate is concentrated under reduced pressure at 30°C. Any remaining traces of solvents are removed by stirring under reduced pressure using an oil pump (20°C/1 mm). This procedure takes about 2 hr and yields ca. 130 g (ca. 1 mol) of the dry, crude N-formyl compound, which is used in the next step without further purification.

C. *(S)-(-)-1-Formyl-2-methoxymethylpyrrolidine*. A 4-L, three-necked flask, fitted with a magnetic stirrer, reflux condenser, low temperature thermometer and a mineral oil bubbler is charged with a solution of the crude formyl derivative in 1.5 L of dry tetrahydrofuran (Note 9) and flushed with argon. The solution is cooled to -50 to -60°C (internal temperature, acetone-dry ice), the cooling bath is removed, and 81 mL (1.3 mol) of methyl iodide is added. Then 28.8 g (1.2 mol) of sodium hydride (Note 10) is introduced carefully in one portion. *CAUTION! To prevent contact of sodium hydride with the cooling medium, it is absolutely necessary to remove the cooling bath before adding the sodium hydride. In addition, argon bubbling is stopped to avoid sodium hydride dust from being blown out of the flask.* The apparatus is flushed again with argon and allowed to warm to room temperature. During this period hydrogen gas evolves and a grey solid precipitates, which causes stirring to become difficult. At about 0°C the precipitate dissolves exothermally under strong evolution of hydrogen. The thermometer is replaced by a pressure-equalizing dropping funnel, the solution is refluxed for 15 min, and quenched by slow addition of 90 mL of 6 N hydrochloric acid, without external heating. Tetrahydrofuran is removed under reduced pressure to yield the crude O-methylated compound in water.

D. *(S)-(+)-2-Methoxymethylpyrrolidine*. A solution of 180 g of potassium hydroxide in 720 mL of water is added to the crude product and the mixture is vigorously stirred under argon overnight. Saturation with potassium carbonate (500 g) causes precipitation of potassium salts, which are filtered off by suction (large Büchner funnel) and washed with ether. The filtrate is extracted with ether (3 x 300 mL) (Note 11) and the ether layer is acidified with 100 mL of 12 N hydrochloric acid under ice cooling in a hood (evolution of fumes) and extracted twice with 100 mL of water to yield an aqueous solution of the hydrochloride of 2-methoxymethylpyrrolidine (Note 12).

E. *(S)-(-)-1-Carbamoyl-2-methoxymethylpyrrolidine*. The aqueous amine hydrochloride solution is adjusted to a pH of 2.8-3.2 (Note 13) with aqueous 50% potassium hydroxide. A solution of 80 g (1 mol) of potassium cyanate (Note 14) in 140 mL of water is then added all at once at 15°C and the mixture is allowed to stir for at least 12 hr at 20°C.

F. *(S)-(-)-1-Amino-2-methoxymethylpyrrolidine (SAMP)*. A 4-L, three-necked flask containing the crude urea is cooled to -5°C (internal temperature) by means of an ice-salt bath and treated with a chilled (-5°C) solution of 168 g of potassium hydroxide in 150 mL of water. After addition of 685 mL (1.3 mol) of 1.9 N potassium hypochlorite solution (Note 15), precooled to -5°C, the temperature rises within 10 min to 30-40°C and the cooling bath is removed after the mixture reaches room temperature (Note 16). Stirring is continued for a total of 12-15 hr. Excess potassium hypochlorite is destroyed with a freshly prepared solution of sodium bisulfite (20 g $NaHSO_3$ in 50 mL of H_2O) and the mixture is acidified (pH 2) with a minimum amount of 12 N hydrochloric acid (ca. 350 mL) under ice cooling. A strong evolution of carbon dioxide occurs. The mixture is allowed to stir for an additional 15 min at ambient temperature, made alkaline (pH 9) with ca. 100 mL of aqueous 50% potassium hydroxide, and saturated with potassium carbonate (500 g). Precipitated potassium salts are filtered off by suction through a large Büchner funnel and washed twice with ethanol; the filtrate is extracted with a 1:1 chloroform/ethanol mixture (1 x 800 mL and 2 x 400 mL). During extraction salt precipitation again occurs and the salts should be filtered off as mentioned above. The organic layers are collected, concentrated under reduced pressure at 30°C (Note 17), taken up in 500 mL of chloroform, and dried twice over sodium sulfate. The drying agent is removed by suction filtration through Celite and the solvent is stripped off as mentioned above to give 80-90 g of a dark oil. Immediate distillation through a 40-cm vacuum-jacketed Vigreux column (the receiver should be cooled with ice to prevent loss of substance) yields a small forerun containing 2-methoxymethylpyrrolidine, followed by SAMP as a colorless liquid, bp 42°C/1.8 mm (80°C bath temperature), $[\alpha]_D^{20}$ -79.6° (neat). Overall yields of SAMP range from 65 to 75 g (50-58%). A purity of ca. 95% was established by GLC analysis (Note 18). The optical antipode RAMP, $[\alpha]_D^{20}$ +79.8° (neat), can be prepared likewise, starting from (R)-proline (Note 19).

2. Notes

1. The heating mantle should be covered with aluminum foil to prevent contact with lithium aluminum hydride and proline.

2. *Peroxide-free* tetrahydrofuran (for precautions, see *Org. Synth., Collect. Vol. V* **1973**, 976) was refluxed over potassium hydroxide pellets for 2 hr, distilled, and dried by addition of ca. 1 g of lithium aluminum hydride prior to use. *Failure to heed the precautions can result in a serious explosion!* Drying the THF over sodium-benzophenone is generally recommended for safety reasons, but this variation was not checked.

3. Lithium aluminum hydride (100%) was used as purchased from Metallgesellschaft AG, Frankfurt, Germany, or Wako Pure Chemical Industries, Ltd., Japan.

4. (S)- and (R)-proline ($\geq$ 99.5% ee) were obtained from Degussa AG, Hanau, Germany. The checkers used (S)-proline (guaranteed reagent) purchased from Wako Pure Chemical Industries, Ltd.

5. A convenient technique is to add the proline portion-wise with a spoon that is inserted as far as possible into the flask to prevent the proline from being blown off the spoon by hydrogen generated during the reaction. The flask is immediately stoppered after every addition.

6. The temperature of the heating bath should not exceed 30°C to avoid significant loss of product. To prevent oxidation the rotary evaporator is flushed with argon.

7. Methyl formate (reagent grade) was used as purchased from Riedel de Haen, Seelze, Germany. The checkers used the product (guaranteed reagent) of Wako Chemical Industries, Ltd.

8. Celite was supplied by Fluka AG, Buchs, Switzerland, or Manville Products Corporation, USA.

9. This time, tetrahydrofuran (Note 2) is made anhydrous by adding ca. 2 g of sodium hydride.

10. Methyl iodide was purchased from Merck-Schuchardt, Hohenbrunn, Germany, or Wako Pure Chemical Industries, Ltd., Japan. Because of its volatility and possible carcinogenicity it should be handled in a well-ventilated hood. Sodium hydride was supplied by Riedel de Haen, Seelze, Germany, or Wako Pure Chemical Industries, Ltd., Japan. In a 1-L, round-bottomed flask 50 g of 80% sodium hydride (mineral oil dispersion) is washed free of oil by stirring with pentane (4 x 300 mL). After the supernatant liquid is decanted for the fourth time, the remaining solvent is removed by evaporation (20°C/1 mm).

11. After the usual workup, the amine can be obtained pure merely by distilling through a 40-cm Vigreux column, bp 75-77°C/40 mm.

12. The ether layer contains a small amount of the starting material and is discarded.

13. The pH of the solution should be exactly calibrated by means of a pH meter in order to prevent side reactions. The use of pH paper is not recommended.

14. Potassium cyanate (technical grade) was supplied by Degussa AG, Hanau, Germany. Alternatively, sodium cyanate may be used. The checkers used the practical grade reagent purchased from Wako Pure Chemical Industries, Ltd., Japan.

15. The optimized preparation of an aqueous potassium hypochlorite solution is a modification of an Organic Syntheses procedure.[2] Two hundred grams of HTH [commercially available as swimming pool sanitizer from Olin Chemicals, 120 Long Ridge Road, Stamford, CT 06904, USA, ca. 68% $Ca(OCl)_2$, or Nippon Soda Co., Japan, 70% $Ca(OCl)_2$] is vigorously shaken with 600 mL of water (ca. 10 min), a solution (20°C) of 40 g of potassium hydroxide and 140 g of potassium carbonate in 250 mL of water is added, and shaking is continued for at least 10 min to yield a semi-fluid gel. The gel is filtered off by suction and thoroughly pressed dry to give 650-770 mL of a yellow potassium hypochlorite solution. These solutions are 1.8-2.2 M according to simple iodometric titration (see *Org. Synth.* **1976**, *56*, 118, Note 3).

16. Lower temperatures than those mentioned above inhibit the exothermic reaction and the reaction time is extended to approximately 24 hr. Less efficient cooling results in warming to 70-80°C, which causes decarboxylation of the intermediate carboxylate and/or simple oxidation by potassium hypochlorite.

17. If the temperature of the water bath exceeds 30°C, significant amounts of product are lost. To prevent oxidation, the rotary evaporator is flushed with argon.

18. The checkers observed contamination of the final product by ca. 4% 2-methoxymethylpyrrolidine according to 270 MHz NMR.

19. The chiral hydrazines are stable over months if stored in a refrigerator under argon. The spectra are as follows. IR (film) cm^{-1}: 3360 (NH_2), 3150, 2980, 2880, 2820, 1610, 1465, 1200, 1120, 960, 920; ^1H NMR ($CDCl_3$, 90 MHz) δ: 1.4-2.1 (m, 4 H, CH_2), 2.1-2.6 (m, 2 H, CH_2N), 3.1 (m, 3 H, NH_2, NCH), 3.3 (s, 3 H, OCH_3), 3.4 (m, 2 H, CH_2O); mass spectrum (70 eV) m/e (rel. intensity): M^+ 130.1100 (6.7%) (calcd. 130.1106); 97.07 (3.9);

86.07 (8.9); 85.07 (100.0); 83.06 (4.1); 71.06 (16.3); 68.05 (31.3); 57.04 (5.6); 56.05 (4.6); 45.03 (10.7); 43.03 (12.1); 42.04 (3.4); 41.04 (28.9); 39.02 (5.5).

3. Discussion

The previously reported preparation of SAMP and its enantiomer RAMP involved hazardous nitrosamine intermediates.[3-5] The new procedure described here circumvents this problem by N-amination via Hofmann-degradation[6] (step F). The procedure, which required an optimization of the synthesis of the known intermediates,[7,8] is characterized by good yields, mild conditions and readily available starting materials. The entire sequence of six steps can be performed in a week.

The enantiomerically pure hydrazines SAMP and RAMP are versatile chiral auxiliaries with a wide range of applications in asymmetric synthesis.[9] For a detailed description of the SAMP/RAMP-hydrazone method see the following Organic Syntheses procedure.

1. Institut für Organische Chemie der Rheinischen Westfälischen Technischen Hochschule, Professor-Pirlet-Strasse 1, 5100 Aachen, Germany.
2. Newman, M. S.; Holmes, H. L. *Org. Synth., Collect. Vol. II* **1943**, 428.
3. Enders, D.; Eichenauer, H. *Chem. Ber.* **1979**, *112*, 2933.
4. Enders, D.; Eichenauer, H.; Pieter, R. *Chem. Ber.* **1979**, *112*, 3703.
5. Enders, D.; Eichenauer, H. *Angew. Chem.* **1976**, *88*, 579; *Angew. Chem., Intern. Ed. Engl.* **1976**, *15*, 549.
6. Sucrow, W. In "Methodicum Chimicum", Korte, F., Ed.; Georg Thieme Verlag, Stuttgart, Academic Press: New York, San Francisco, London, 1975; Vol. 6, pp. 99-104.

7. Doyle, F. P.; Mehta, M. D.; Sach, G. S.; Pearson, J. L. *J. Chem. Soc.* **1958**, 4458.

8. Seebach, D.; Kalinowski, H.-O.; Bastani, B.; Crass, G.; Daum, H.; Dörr, H.; Du Preez, N. P.; Ehrig, V.; Langer, W.; Nüssler, Ch.; Oei, H.-A.; Schmidt, M. *Helv. Chim. Acta* **1977**, *60*, 301.

9. Enders, D. "Alkylation of Chiral Hydrazones" In "Asymmetric Synthesis", Morrison, J. D., Ed.; Academic Press: New York, 1984; Vol. 3, pp. 275.

Appendix
Chemical Abstracts Nomenclature (Collective Index Number);
(Registry Number)

(S)-(-)-1-Amino-2-methoxymethylpyrrolidine (SAMP): 1-Pyrrolidinamine, 2-(methoxymethyl)-, (S)- (9); (59983-39-0)

(R)-(+)-1-Amino-2-methoxymethylpyrrolidine (RAMP): 1-Pyrrolidinamine, 2-(methoxymethyl)-, (R)- (9): (72748-99-3)

L-Proline (8,9); (147-85-3)

D-Proline (8,9); (344-25-2)

(S)-(+)-2-Hydroxymethylpyrrolidine: 2-Pyrrolidinemethanol, (S)-(+)- (8,9); (23356-96-9)

(S)-(-)-1-Formyl-2-hydroxymethylpyrrolidine: 1-Pyrrolidinecarboxaldehyde, 2-(hydroxymethyl)-, (S)- (9); (55456-46-7)

(S)-(-)-1-Formyl-2-methoxymethylpyrrolidine: 1-Pyrrolidinecarboxaldehyde, 2-(methoxymethyl)-, (S)- (10); (63126-45-4)

(S)-(+)-2-Methoxymethylpyrrolidine: Pyrrolidine, 2-(methoxymethyl)-, (S)- (10); (63126-47-6)

ASYMMETRIC SYNTHESES USING THE SAMP-/RAMP-HYDRAZONE METHOD: (S)-(+)-4-METHYL-3-HEPTANONE

(3-Heptanone, 4-methyl, (S)-)

Submitted by Dieter Enders, Helmut Kipphardt, and Peter Fey.[1]
Checked by Benjamin Guzmán, Stan S. Hall, and Gabriel Saucy.

1. Procedure

A. *3-Pentanone SAMP hydrazone [(S)-2].* A 50-mL, one-necked, pear-shaped flask equipped with a 10-cm Liebig condenser, gas inlet tube, and a magnetic stirring bar is charged with 3.9 g (30 mmol) of SAMP (Note 1) and 3.79 mL (36 mmol) of 3-pentanone (Note 2) and the mixture is warmed at 60°C under argon overnight (Note 3). The crude product is diluted with 200 mL of ether in a 250-mL separatory funnel and washed with 30 mL of water. The organic layer is separated, dried over anhydrous magnesium sulfate and concentrated under reduced pressure. Purification by short-path distillation yields 5.18 g (87%) of a colorless oil, bp 70-75°C/0.5 mm, $[\alpha]_D^{20}$ +297° (benzene, c = 1). The SAMP-hydrazone (S)-2 should be stored in a refrigerator under argon (Note 4).

B. *(S)-(+)-4-Methyl-3-heptanone SAMP hydrazone [(2SS)-3].* A flame dried, one-necked, 250-mL flask with side arm, rubber septum and magnetic stirring bar is flushed with argon (Note 5). The flask is then cooled to 0°C and 110 mL of dry ether (Note 6) and 2.97 mL (21 mmol) of dry diisopropylamine (Note 7) are added, followed by dropwise addition of 21 mmol of butyllithium (13.1 mL of a 1.6 N solution in hexane, Note 8). Stirring is continued for 10 min and a solution of 3.96 g (20 mmol) of SAMP-hydrazone (S)-2 in 10 mL of ether is added to the stirred mixture over a period of 5 min at 0°C. Additional 2 mL of ether are used to transfer all of the hydrazone (S)-2 into the reaction flask. Stirring is continued for 4 hr at 0°C, while the lithiated hydrazone precipitates. The mixture is cooled to -110°C (pentane/liquid nitrogen bath) and kept for 15 min at this temperature. Then 2.15 mL (22 mmol) of propyl iodide (Note 9) is added dropwise, and the mixture is allowed to reach room temperature overnight. The contents of the flask are poured into a mixture of 300 mL of ether-50 mL of water in a 500-mL separatory funnel, the layers are separated, and the aqueous layer is extracted twice with 25 mL of ether. The combined organic layers are washed with 10 mL of water, dried over anhydrous magnesium sulfate, and concentrated under reduced pressure to yield 4.3 g (90%) of crude (2S)-3 (Note 10).

C. *(S)-(+)-4-Methyl-3-heptanone [(S)-4].* A 100-mL Schlenk tube, fitted with a gas inlet and Teflon stopcocks, is charged with 4.3 g (18 mmol) of crude (2S)-3 dissolved in 50 mL of dichloromethane (Note 11), and cooled to -78°C (acetone/dry ice bath) under nitrogen. Dry ozone (Note 12) is passed through the yellow solution until a green-blue color appears (ca. 4 hr). The mixture is then allowed to come to room temperature while a stream of nitrogen is bubbled through the solution to give the yellow nitrosamine (S)-**5** (Note 13) and the title ketone (S)-**4**. The solvent is removed by distillation at 760 mm (60°C bath temperature) and the residue is transferred into a microdistillation apparatus (10-mL flask, 5-cm Vigreux column, spider, collection device, Note 14). After a small forerun (3-4 pale yellow drops), a colorless liquid distills to afford 1.6-1.7 g (56-58% overall) of ketone (S)-**4**; bp 63-67°C/40 mm (110-115°C bath temperature), GLC analysis 98.2%, $[\alpha]_D^{20}$ +21.4° to +21.7° (hexane, c = 2.2), $[\alpha]_D^{20}$ +17.88° (neat) (Note 15). To recycle the chiral auxiliary SAMP see Note 16.

2. Notes

1. (S)-1-Amino-2-methoxymethylpyrrolidine (SAMP) and RAMP are commercially available from Merck-Schuchardt (Frankfurter Strasse 250, 6100 Darmstadt or Eduard-Buchner-Strasse 14-20, 8011 Hohenbrunn, Germany) and Aldrich Chemical Company, Inc., Milwaukee, Wisconsin. The submitters prepared SAMP from (S)-proline as described in the accompanying procedure; the checkers used SAMP from Merck-Schuchardt.

2. Redistilled prior to use, 3-pentanone was obtained from Merck-Schuchardt (submitters) and Aldrich Chemical Company, Inc. (checkers).

3. The reaction was monitored by TLC. The TLC plates (SiO_2, F_{254}, 0.25 mm), commercially available from Merck, Darmstadt, Germany, were eluted with ether and developed by dipping into a 10% ethanolic solution of phosphomolybdic acid (Merck) and then heating.[2]

4. Partial ^{1}H NMR spectrum (CDCl$_3$, 200 MHz) δ: 1.06 (t, 3 H, J = 7.7, C$\underline{H}_3$CH$_2$), 1.08 (t, 3 H, J = 7.5, C$\underline{H}_3$CH$_2$), 3.34 (s, 3 H, C$\underline{H}_3$O); IR (film) cm^{-1}: 1645.

5. This is done by alternately evacuating and filling the flask with argon three times. During the reaction a pressure of about 50 mm above atmospheric pressure is maintained using a mercury bubbler. All reagents are added via a glass syringe under rigorously anhydrous conditions. For a more detailed description of the metalation technique see reference 3.

6. The submitters used ether that had been freshly distilled from sodium and benzophenone under an argon atmosphere. The checkers used anhydrous ether directly from freshly opened 500-g containers from Fisher Scientific Company, Springfield, New Jersey. In addition, the checkers charged the reaction flask with ether using a dry graduated cylinder, flushed the system with argon and then sealed and cooled the vessel to 0°C before the sequential addition of diisopropylamine and butyllithium.

7. Diisopropylamine from BASF AG, Ludwigshafen, Germany (submitters) and Aldrich Chemical Company, Inc. (checkers) was distilled from calcium hydride, and then stored under argon and over calcium hydride until use.

8. Butyllithium was purchased from Metallgesellschaft, Frankfurt, Germany, and titrated for active alkyllithium using diphenylacetic acid as an indicator.[4] The checkers used fresh butyllithium, 1.55 M in hexane under argon, from Aldrich Chemical Company, Inc. and omitted the titration.

9. Propyl iodide was obtained from Merck-Schuchardt, distilled over potassium carbonate and stored over copper wire under argon. *Caution: Propyl iodide is a cancer suspect agent.* The checker's source was Aldrich Chemical Company, Inc.

10. Partial ^{1}H NMR spectrum (CDCl$_3$, 200 MHz) δ: 0.88 (t, 3 H, J = 6.9, C$\underline{H}_3$CH$_2$), 1.02 (d, 3 H, J = 7.1, C$\underline{H}_3$CH), 1.10 (t, 3 H, J = 7.5, C$\underline{H}_3$CH$_2$), 3.33 (s, 3 H, C$\underline{H}_3$O); IR (film) cm^{-1}: 1630. An ^{1}H-NMR experiment using the chiral shift reagent [Eu(hfc)$_3$, Aldrich] with crude 3 shows that only the (ZSS) isomer is present (sharp methoxy singlet). During the measurement a slow isomerization to the thermodynamically more stable (ESS) isomer takes place, but within the limit of detection of a 100-MHz spectrometer no (SR) diastereomer can be seen (diastereomeric excess de >97%).[5]

11. Dichloromethane was distilled over potassium carbonate prior to use.

12. *Caution! Organic ozonides are highly explosive. The reaction should be carried out in a well-ventilated hood with a shatter-proof shield. Do not grease the ground glass joints!* The submitters used a Fischer Model OZ II ozonizer from Fischer, Bad Godesberg, Germany. For detailed descriptions of a laboratory ozonizer see *Org. Synth., Collect. Vol. III* **1955**, 673. The checkers used a Welsbach Model T-408 Laboratory Ozonizer, Welsbach Corp., Philadelphia, PA. The power setting was 100 volts (AC) and the oxygen pressure setting was 8 psi (0.55 Kg/cm^2) to produce 2-3% ozone at a gas flow rate (rotameter) of 2 L/min. The ozone production rate was measured by passing a measured amount of ozonized gas through a 2% potassium iodide solution (neutral), acidifying with 1 M sulfuric acid, and then titrating the liberated iodine with 0.1 N sodium thiosulfate. Using these conditions, the ozonolysis required at least 4 hr, rather than the 30 min suggested by the submitters.

13. *Caution! The nitrosamine (S)-5 may be carcinogenic. All operations with (S)-5 should be performed in a well-ventilated hood, and the operator should wear disposable gloves.* In order to destroy any nitrosamine traces, the glassware contaminated with (S)-5 should be immersed in a bath of HBr/acetic acid.

14. To prevent any racemization during distillation, the apparatus is shaken with 1 mL of chlorotrimethylsilane (freshly distilled from calcium hydride), which is removed under reduced pressure. *Caution: glassware, cleaned under alkaline conditions, will lead to spontaneous racemization!* The spider (3-4 arms for liquid collection) should be cooled with an ice bath. To prevent bumping, the checkers performed the distillation with a Bunsen burner rather than with a bath.

15. The product has an optical purity of 97-98% by comparison with the reported optical rotation of $[\alpha]_D^{25}$ +22.1 ± 0.4° (hexane, c = 1.0) of the naturally occurring pheromone[6] and an ee of ≥ 97% by comparison with the de of ≥ 97% of (2S)-3 (Note 10); ^{1}H NMR (CDCl$_3$, 200 MHz) δ: 0.90 (t, 3 H, J = 6.7, C$\underline{H}_3$CH$_2$), 1.06 (d, 3 H, J = 6.9, C$\underline{H}_3$CH) superimposed on 1.04 (t, 3 H, J = 7.2, C$\underline{H}_3$CH$_2$), 2.45 (q, 2 H, J = 7.3, C$\underline{H}_2$CH$_3$); IR (film) cm^{-1}: 1710.

16. The nitrosamine (S)-5 (1.94 g, 67%) is obtained from the residue of the ketone distillation (bp 79-80°C/0.1 mm). Reduction with lithium aluminum hydride in tetrahydrofuran yields 1.47 g (49% overall) of SAMP[7] with an $[\alpha]_D^{20}$ -75.46° (neat).

3. Discussion

The title ketone (S)-**4**, which is 400 times more active than its optical antipode,[6,8] is the principal alarm pheromone of the leaf cutting ant *Atta texana*. (S)-**4** has also been identified as an alarm pheromone in three other ant genera of the subfamily *Myrmicinae*,[6,9] as a component of the defensive secretion of the "daddy longleg" *Leiobunum vittatum* (Opiliones),[10,11] and is produced by the elm bark beetles *Scolytus scolytus (F.)* and *S.multistriatus*.[12]

(S)-**4** and/or its enantiomer (R)-**4** have been prepared via resolution of an intermediate,[8] starting from (R)-citronellic acid,[13] by stoichiometric asymmetric synthesis[14-16] (76-88% ee), and by a microbiological method.[17]

The three-step procedure described here, using inexpensive, commercially available starting materials and the chiral auxiliary SAMP, illustrates the synthetic utility of the "SAMP-/RAMP-hydrazone method".[18] It is remarkable that the classical electrophilic substitution of a conformationally flexible, acyclic ketone **1** → (S)-**4** occurs with virtually complete asymmetric induction. This demonstrates complete stereochemical control of the three critical operations: metalation, alkylation, and cleavage. Because deprotonated SAMP-/RAMP-hydrazones react with nearly the entire palette of electrophiles, this new methodology, a chiral version of the now widely used dimethylhydrazone (DMH) method,[3] opens an elegant and economical entry to a great variety of important classes of compounds with good overall chemical yields and excellent diastereo- and enantioselectivities. The following stereoselective reactions may be mentioned: α-alkylations of aldehydes[7a,19] and ketones,[5,7a,11] diastereo- and enantioselective aldol reactions,[7b,20,21] diastereo- and enantioselective Michael additions to form β,γ-substituted δ-keto esters,[22,23] δ-lactones,[24] and various heterocycles, such as dihydropyridines, octahyroquinolinediones and hexahydroquinolinones,[25] α-alkylations of β-keto esters,[18] and, finally, asymmetric syntheses of α- and/or β-substituted primary amines[18c,d] via alkylation/reductive amination of aldehydes and ketones[26] or nucleophilic addition to aldehyde-SAMP-/RAMP-hydrazones, followed by N-N bond cleavage.[27] This broad applicability is summarized in Fig. 1 and typical examples are listed in Table 1.

In S_E2'-front type electrophilic substitutions via SAMP-/RAMP-hydrazones the less substituted α carbon atom is regioselectively deprotonated. Under the standard reaction conditions (lithium diisopropylamide, 0°C, ether or THF) the intermediate aza enolates are formed as the $E_{CC}Z_{CN}$ species, as confirmed by spectroscopic[19,28] and numerous trapping experiments.[18] Because of the uniform diastereoface differentiation common for all asymmetric SAMP-/RAMP-hydrazone alkylations, the absolute configuration that will predominate in the final product can be predicted reliably. Furthermore, instead of changing from SAMP to the enantomeric RAMP as chiral auxiliary, it is possible to prepare both enantiomers of target molecules in excess using SAMP, simply by changing the building blocks used as nucleophile and electrophile. This opposite enantioselectivity through synthon control is demonstrated in the cases of 2-methylbutanal, 2-methyloctanal, and 2-methyl-1-octanamine (see Table 1).

Another advantage of SAMP-/RAMP-hydrazones is the facile determination of the asymmetric induction by downfield shifting of the SAMP- or RAMP-hydrazone methoxy singlet [LIS-technique, $Eu(fod)_3$].

Besides the oxidative cleavage by ozonolysis, the optically active carbonyl compounds can be alternatively obtained by acidic hydrolysis of the corresponding SAMP-/RAMP-hydrazone methiodides in a two-phase system.[5b,7a]

The chiral auxiliary SAMP or RAMP may be recycled by lithium aluminum hydride-reduction of the nitrosamine (S)-5 formed during ozonolysis. Other very successful applications of the SAMP-/RAMP-hydrazone method in natural product synthesis have recently been reported by Nicolaou, et al. [ionophore antibiotic X-14547A (indanomycine)];[29] Pennanen (eremophilenolide, sesquiterpene);[30] Enders, et al. (defensive substance of "daddy longlegs");[11] Mori, et al. (serricornin, cigarette beetle pheromone);[31] and Bestmann, et al. (pheromone analogues).[32] Finally, it should be mentioned that the chiral auxiliaries SAMP and RAMP may also be used in the resolution of aldehydes[33] and ketones,[34] and in the NMR spectroscopic determination of % ee of chiral aldehydes.[35]

Figure 1. Optically Active Carbonyl Compounds and Amines

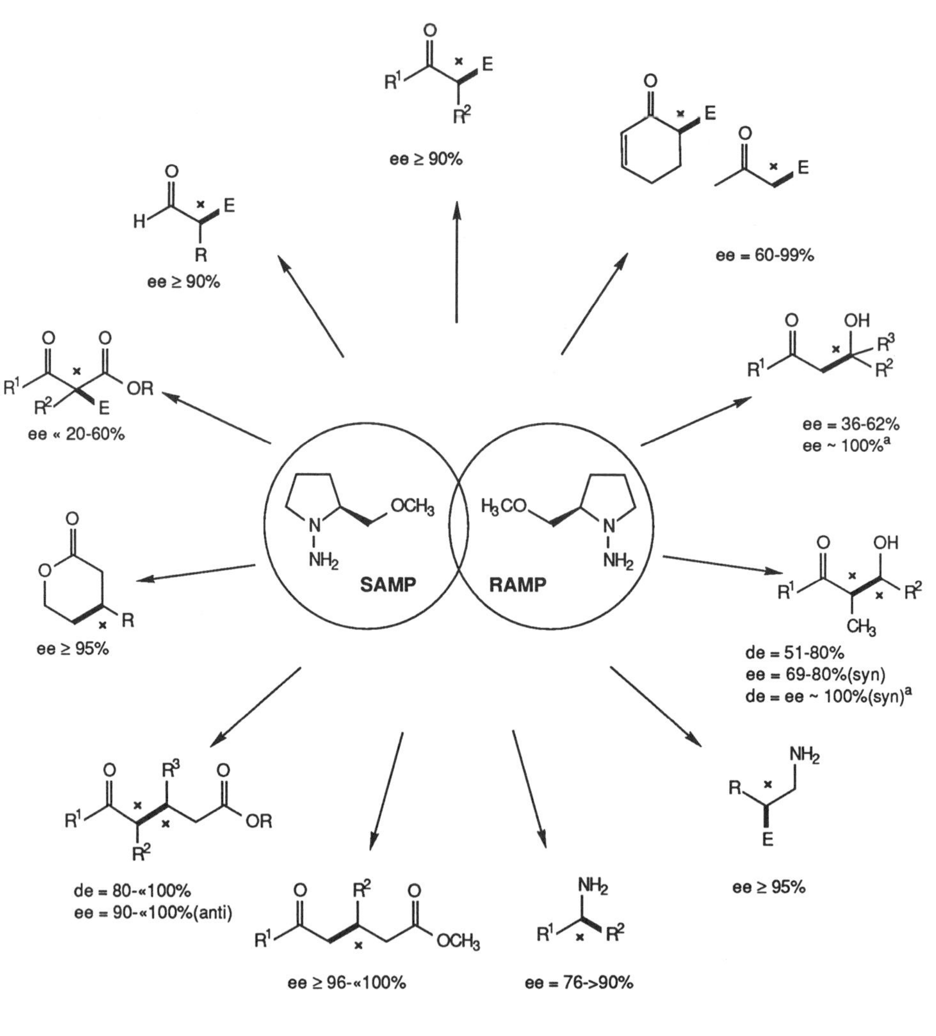

[a] After crystallization

Table I. Optically Active Carbonyl Compounds and Amines Prepared by Asymmetric Synthesis Using the SAMP/RAMP Hydrazone Method

Carbonyl Compound or Amine	Electrophile	Cleavage[a]	Yield[b] [%]	ee[%] (Config.)	Lit.
H_3C-CH(CH$_3$)-CHO	C_2H_5I	B	71	95 (S)	7a,18a
H_3C-CH(CH$_3$)-CHO	CH_3I	A	65	95 (R)	7a,18a
$H_3C(CH_2)_5$-CH(CH$_3$)-CHO	$C_6H_{13}I$	A	52	≥95 (S)	7a,18a
$H_3C(CH_2)_5$-CH(CH$_3$)-CHO	CH_3I	A	61	95 (R)	7a,18a
$CH_2=CH(CH_2)_7$-CH(CH$_3$)-CHO	$(CH_3)_2SO_4$	B	65	95 (R)	18a,32
$CH_2=CH(CH_2)_7$-CH(CH$_3$)-CHO	$(CH_3)_2SO_4$	B	51	95(S)[c]	18a,32
2-methylcyclopentanone	$(CH_3)_2SO_4$	A	66	86 (R)	7a
	CH_3I	A	74	45 (R)	7a

193

Table I. Continued

Carbonyl Compound or Amine	Electrophile	Cleavage[a]	Yield[b] [%]	ee[%] (Config.)	Lit.
2-methylcyclohexanone	(CH$_3$)$_2$SO$_4$	A	70	»99 (R)	7a
	CH$_3$I	A	70	67 (R)	7a
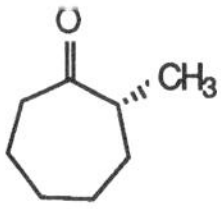 2-methylcycloheptanone	CH$_3$I	B	59	94 (R)	7a
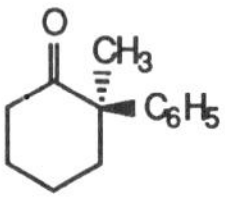 2-methyl-2-phenylcyclohexanone	CH$_3$I	C	43	93 (R)	18a,c
6-(but-3-enyl)cyclohex-2-enone	H$_2$C=CH(CH$_2$)$_2$Br	B	26	≥89 (R)	18a,30
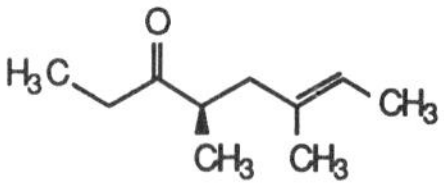 H$_3$C ketone with CH$_3$ CH$_3$ and =CH$_3$	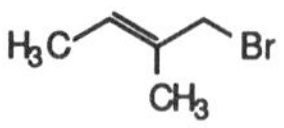 H$_3$C—CH$_2$Br with CH$_3$	B	61	≥97 (S)	11
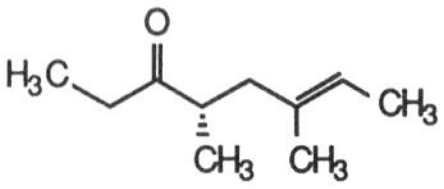 H$_3$C ketone with CH$_3$ CH$_3$ and =CH$_3$	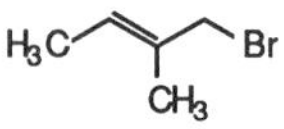 H$_3$C—CH$_2$Br with CH$_3$	B	62	≥97 (R)[c]	11
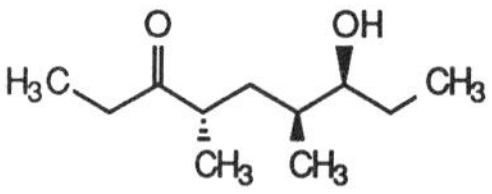 H$_3$C ketone with OH and CH$_3$	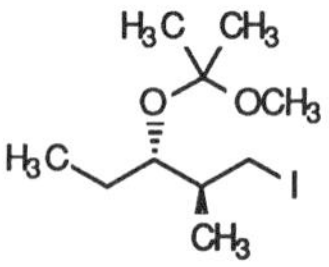 H$_3$C CH$_3$ / O-C-OCH$_3$ / H$_3$C...CH$_2$I with CH$_3$	B	46	99 (S)	31

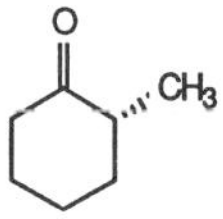

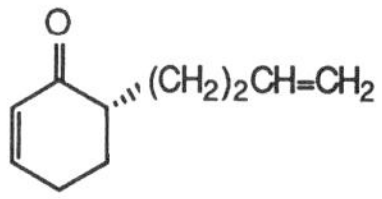

194

Table I. Continued

Carbonyl Compound or Amine	Electrophile	Cleavage[a]	Yield[b] [%]	ee[%] (Config.)	Lit.
H₃C-CO-CH(CH₃)-CH₂-C(O)-Ot-C₄H₉	t-C₄H₉O-C(O)-CH₂Br	A	53	>95 (S)	18a,c
H₃C-CO-CH(CH₃)-CH₂-C(OC₂H₅)=CH-CO₂C₂H₅	BrCH₂-C(OC₂H₅)=CH-CO₂C₂H₅	A	80	≥95 (S)	18a,c
C₆H₅-CO-CH(CH₃)-CH₂-CH₃	C₂H₅I	B	44	≥97 (S)	5b
t-C₄H₉-CO-CH₂-CH(OH)-c-C₆H₁₁	c-C₆H₁₁CHO	D	32	62 (−) »100 (−)[e]	20
H₃CO-(HO-C₆H₃)-CH₂-CH₂-CO-CH₂-CH(OH)-C₅H₁₁	C₅H₁₁CHO	E	75	39 (S)[c]	7b
C₆H₅-CO-CH(CH₃)-CH(OH)-C₆H₅	C₆H₅CHO	A	86	72 (SS) [de=67% (SS)]	21
		A	35	de=ee≈100 (SS)[f]	
H₃C-CO-CH(CH₃)-CH(OH)-C₆H₅	C₆H₅CHO	A	81	74 (SS) [de=66% (SS)]	21
		A	37	de=ee≈100 (SS)[f]	

Table I. Continued

Carbonyl Compound or Amine	Electrophile	Cleavage[a]	Yield[b] [%]	ee[%] (Config.)	Lit.
(structure)	(structure)	A	62	≥99 (R)	22
(structure)	(structure)	A	45	≥96 (R)	22
(structure)	(structure)	A	46	≥95 (R)	24
(structure)	(structure)	A	39	≥98 (SS) [de ~ 100% (SS)]	23
(structure)	(structure)	A	35[g]	≥95 (R)	24
(structure)	(structure)	A	30[g]	≥95 (R)	24
(structure)	CH₃I	A	65	60 (-)	18a,c

Table I. Continued

Carbonyl Compound or Amine	Electrophile	Cleavage[a]	Yield[b] [%]	ee[%] (Config.)	Lit.
H₃C-CO-C*(CH₃)(C₂H₅)-CO-OC₂H₅	C₂H₅I	A	52	27(+)	18a,c
C₆H₁₃-CH(CH₃)-CH₂-NH₂	CH₃I	h	56	>90 (R)	26
C₆H₁₃-CH(CH₃)-CH₂-NH₂	C₆H₁₃Br	h	63	>95 (S)	26
t-C₄H₉-CH(CH₃)-NH₂	-	i	60	90 (S)	18c,d
t-C₄H₉-CH(CH₃)-NH₂	-	j	47	88 (R)	18c,d
C₆H₅-CH((CH₂)₃CH₃)-NH₂	-	k	56	85 (R)	27
t-C₄H₉-CH(CH₃)-NH₂	-	l	41	87 (R)	27

[a]A: Oxidative cleavage by ozonolysis (O_3, CH_2Cl_2, -78°C). B: Acidic hydrolysis (i. excess MeI, 60°C, ii. 5 N HCl/pentane). C: Acidic hydrolysis (12 N HCl/ether). D: Oxidative cleavage ($^1O_2$, Me_2S, hydrolysis). E: Oxidative cleavage (30% H_2O_2, MeOH, pH 7 buffer).

[b]Overall chemical yield.

[c]RAMP was used as chiral auxiliary.

[d]de of corresponding SAMP-hydrazone.

[e]After two recrystallizations of the ketol.

[f]After separation and cleavage of the corresponding crystalline SAMP-hydrazone.

[g]Overall yield, including reduction of the intermediate β-substituted aldehyde esters and lactonization.

[h]The primary amines are obtained by catechol-borane reduction of the SAMP-hydrazones, followed by N-N bond cleavage with Raney Nickel.

[i]Obtained by $LiAlH_4$-reduction of 3,3-dimethyl-2-butanone-SAMP-hydrazone, followed by N-N bond cleavage.

[j]Obtained by catechol-borane reduction of 3,3-dimethyl-2-butanone-SAMP-hydrazone, followed by N-N bond cleavage.

[k]Obtained by addition of butyllithium to benzaldehyde-SAMP-hydrazone, followed by N-N bond cleavage.

[l]Obtained by addition of methyllithium to 2,2-dimethylpropanal-SAMP-hydrazone, followed by N-N bond cleavage.

1. Institut für Organische Chemie der Rheinischen Westfälischen Technischen Hochschule, Professor-Pirlet-Strasse 1, 5100 Aachen, Germany.
2. Enders, D.; Pieter, R. *Chem. Labor Betr.* **1977**, *28*, 503; *Chem. Abstr.* **1978**, *88*, 145652r.
3. Corey, E. J.; Enders, D. *Chem. Ber.* **1978**, *111*, 1337, 1362.
4. Kofron, W. G.; Backlawski, L. M. *J. Org. Chem.* **1976**, *41*, 1879.
5. (a) Enders, D.; Eichenauer, H. *Angew. Chem.* **1979**, *91*, 425; *Angew. Chem., Inter. Ed. Engl.* **1979**, *18*, 397; (b) Enders, D.; Eichenauer, H.; Baus, U.; Schubert, H.; Kremer, K. A. M. *Tetrahedron* **1984**, *40*, 1345.
6. Riley, R. G.; Silverstein, R. M.; Moser, J. C. *Science* **1974**, *183*, 760.
7. (a) Enders, D.; Eichenauer, H. *Chem. Ber.* **1979**, *112*, 2933; (b) Enders, D.; Eichenauer, H.; Pieter, R. *Chem. Ber.* **1979**, *112*, 3703.
8. Riley, R. G.; Silverstein, R. M. *Tetrahedron* **1974**, *30*, 1171.
9. Riley, R. G.; Silverstein, R. M.; Moser, J. C. *J. Insect. Physiol.* **1974**, *20*, 1629; *Chem. Abstr.* **1974**, *81*, 133106h; Blum, M. S.; Padovani, F.; Amante, E. *Comp. Biochem. Physiol.* **1968**, *26*, 291; *Chem. Abstr.* **1968**, *69*, 41984s; McGurk, D. J.; Frost, J.; Eisenbaum, E. J.; Vick, K.; Drew, W. A.; Young, J. *J. Insect. Physiol.* **1966**, *12*, 1435; *Chem. Abstr.* **1967**, *66*, 26859z.
10. Meinwald, J.; Kluge, A. F.; Carrel, J. E.; Eisner, T. *Proc. Natl. Acad. Sci. U.S.A.* **1971**, *68*, 1467.
11. Enders, D.; Baus, U. *Liebigs Ann. Chem.* **1983**, 1439.
12. Blight, M. M.; Henderson, N. C.; Wadhams, L. J. *Insect. Biochem.* **1983**, *13*, 27; *Chem. Abstr.* **1983**, *98*, 104645d.
13. Mori, K. *Tetrahedron* **1977**, *33*, 289.

14. Hegedus, L. S.; Kendall, P. M.; Lo, S. M.; Sheats, J. R. *J. Am. Chem. Soc.* **1975**, *97*, 5448.
15. Larcheveque, M.; Ignatova, E.; Cuvigny, T. *J. Organomet. Chem.* **1979**, *177*, 5.
16. (a) Meyers, A. I.; Williams, D. R.; White, S.; Erickson, G. W. *J. Am. Chem. Soc.* **1981**, *103*, 3088; (b) Rossi, R.; Marasco, M. *Chim. Ind. (Milan)* **1980**, *62*, 314; *Chem. Abstr.* **1981**, *94*, 15143p.
17. Kergomard, A.; Renard, M. F.; Veschambre, H. *J. Org. Chem.* **1982**, *47*, 792.
18. For reviews see: (a) Enders, D. "Alkylation of Chiral Hydrazones" in "Asymmetric Synthesis", Morrison, J. R., Ed.; Academic Press: New York, 1984; Vol. 3, pp. 275; (b) Enders, D. *CHEMTECH* **1981**, *11*, 504; (c) Enders, D. "Regio-, Diastereo-, and Enantioselective C-C Coupling Reactions Using Metalated Hydrazones, Formamides, Allylamines, and Aminonitriles" in "Current Trends in Organic Synthesis", Nozaki, H., Ed.; Pergamon Press: Oxford, 1983; p. 151; (d) Enders, D. "Asymmetric Synthesis of Carbonyl Compounds and Primary Amines" in "Selectivity - a Goal for Synthetic Efficiency", Bartmann, W.; Trost, B. M., Eds.; Verlag Chemie: Weinheim, 1984; p. 65. (e) Enders, D. *Chemica Scripta* **1985**, *25*, 139.
19. Davenport, K. G.; Eichenauer, H.; Enders, D.; Newcomb, M.; Bergbreiter, D E. *J. Am. Chem. Soc.* **1979**, *101*, 5654.
20. Eichenauer, H.; Friedrich, E.; Lutz, W.; Enders, D. *Angew. Chem.* **1978**, *90*, 219; *Angew. Chem., Inter. Ed. Engl.* **1978**, *17*, 206.
21. Enders, D.; Baus, U.; Kremer, K. A. M. *Angew. Chem.* **1986**, *98*, to be published.
22. Enders, D.; Papadopoulos, K. *Tetrahedron Lett.* **1983**, *24*, 4967.
23. Enders, D.; Papadopoulos, K.; Rendenbach, B. E. M.; Appel, R.; Knoch, F. *Tetrahedron Lett.* **1986**, *22*, in press.

24. Enders, D.; Rendenbach, B. E. M. to be published.
25. Demir, A. S. Dissertation, University of Bonn, 1985.
26. Enders, D.; Schubert, H. *Angew. Chem.* **1984**, *96*, 368; *Angew. Chem., Inter. Ed. Engl.* **1984**, *23*, 365; Schubert, H. Dissertation, University of Bonn, 1985.
27. Enders, D.; Schubert, H.; Nübling, C. *Angew. Chem.* **1986**, *98*, to be published.
28. (a) Ahlbrecht, H.; Düber, E. O.; Enders, D.; Eichenauer, H.; Weuster, P. *Tetrahedron Lett.* **1978**, 3691; (b) Rademacher, P.; Pfeffer, H.-U.; Enders, D.; Eichenauer, H.; Weuster, P. *J. Chem. Res. (S)* **1979**, 222; (M) **1979**, 2501.
29. (a) Nicolaou, K. C.; Papahatjis, D. P.; Claremon, D. A.; Dolle, R. E., III *J. Am. Chem. Soc.* **1981**, *103*, 6967; (b) Nicolaou, K. C.; Claremon, D. A.; Papahatjis, D. P.; Magolda, R. L. *J. Am. Chem. Soc.* **1981**, *103*, 6969.
30. Pennanen, S. I. *Acta Chem. Scand.* **1981**, *B35*, 555.
31. Mori, K.; Nomi, H.; Chuman, T.; Kohno, M.; Kato, K.; Noguchi, M. *Tetrahedron* **1982**, *38*, 3705.
32. Bestmann, H. J.; Hirsch, M. L. private communication; Hirsch, M. L. Dissertation, University of Erlangen, 1982.
33. Mies, W. Dissertation, University of Bonn, 1985.
34. Dominguez, D.; Ardecky, R. J.; Cava, M. P. *J. Am. Chem. Soc.* **1983**, *105*, 1608.
35. Enders, D.; Rüb, L.; Breitmaier, E. *Can. J. Chem.* **1986**, to be published.

Appendix

Chemical Abstracts Nomenclature (Collective Index Number);

(Registry Number)

(S)-(+)-4-Methyl-3-heptanone: 3-Heptanone, 4-methyl-, (S)- (9); (51532-30-0)

3-Pentanone SAMP-hydrazone: 1-Pyrrolidinamine, N-(1-ethylpropylidene)-2-(methoxymethyl)-, (S)- (9); (59983-36-7)

SAMP: (S)-1-Amino-2-methoxymethylpyrrolidine: 1-Pyrrolidinamine, 2-(methoxymethyl)-, (S)- (9); (59983-39-0)

3-Pentanone (8,9); (96-22-0)

(S)-(+)-4-Methyl-3-heptanone SAMP-hydrazone: 1-Pyrrolidinamine, N-(1-ethyl-2-methylpentylidene)-2-(methoxymethyl)-, [S-[R*,R*-(Z)]- (10); (69943-24-4)

Propyl iodide: Propane, 1-iodo- (8,9); (107-08-4)

(S)-1-Nitroso-2-methoxymethylpyrrolidine: Pyrrolidine, 2-(methoxymethyl)-1-nitroso-, (S-) (9); (60096-50-6)

(-)-8-PHENYLMENTHOL

(Cyclohexanol, 5-Methyl-2-(1-methyl-1-phenylethyl)-, [1R-(1α,2β,5α)]-)

A. 1 (R)-pulegone → (1. PhMgBr, CuBr, Et$_2$O, -20°C; 2. 2N HCl; 3. KOH/EtOH reflux) → trans-2 + cis-2 ~87 : 13

B. trans-2 + cis-2 → (Na, iPrOH, toluene, reflux) → 3 + 4 ~87 : 13

C. 3 + 4 → (ClCH$_2$COCl, C$_6$H$_5$N(CH$_3$)$_2$, Et$_2$O, 0°C → reflux) → 5

D. 5 → (KOH, EtOH, reflux) → 3

Submitted by Oswald Ort.[1]

Checked by Lalith R. Jayasinghe and James D. White.

1. Procedure

Caution! Chloroacetyl chloride is a strong lachrymator. N,N-Dimethylaniline is a severe poison. Synthetic work with these substances should be performed in an efficient hood.

A. **(2RS,5R)-5-Methyl-2-(1-methyl-1-phenylethyl)cyclohexanone 2** (Note 1). a) Grignard reagent formation and conjugate addition. In a nitrogen-flushed, 500-mL, two-necked, round-bottomed reaction flask fitted with a reflux condenser carrying a calcium chloride tube, 250-mL pressure-equalizing dropping funnel and Teflon coated magnetic stirring bar are placed 11.0 g (0.45 mol) of magnesium turnings and 50 mL of diethyl ether (Note 2). To this flask is added 1/10 of 78.5 g (0.5 mol) of bromobenzene in one portion (Note 3). The reaction mixture is heated to reflux without stirring to start Grignard reagent formation. When the reaction has started (Note 4), the rest of the bromobenzene in 100 mL of diethyl ether is added with stirring at such a rate that gentle reflux is maintained. After the addition is complete, the reaction mixture is heated to reflux for an additional 1 hr. The solution is cooled to room temperature and diethyl ether is added to give a total volume of about 300 mL (Note 5). The reflux condenser and the dropping funnel are replaced by a nitrogen inlet tube and a pierced rubber septum with Teflon tube inlet (Note 6).

In a second nitrogen-flushed, 500-mL, three-necked, round-bottomed reaction flask with a mechanical stirrer, thermometer, and two-way adapter carrying a calcium chloride tube and a rubber septum with Teflon tube, connected to reaction flask 1 (Figure I), are placed 4.4 g (31 mmol) of copper(I) bromide (Note 7) and 70 mL of diethyl ether. The ethereal Grignard solution from reaction flask 1 is added, through the Teflon tube by means of nitrogen pressure (see Figure I), to this vigorously stirred suspension at -20°C. After the addition is complete, the reaction mixture is stirred at -20°C for 1/2 hr. The rubber septum is replaced by a 100-mL, pressure-equalizing dropping funnel containing 40.0 g (0.26 mol) of (R)-(+)-pulegone (Note 8) in 50 mL of diethyl ether. This solution is added with stirring at -20°C to the dark-green reaction mixture during about 2 hr. After the reaction mixture is kept overnight at -20°C, it is added to 300 mL of vigorously stirred ice-cold 2 N hydrochloric acid. The organic layer is separated, filtered with suction and the residue on the funnel is washed twice with 20-mL portions of ether. The aqueous layer is saturated with ammonium chloride and extracted three times with 100-mL portions of ether. The combined organic phases are washed with saturated aqueous sodium hydrogen carbonate solution and the solvent is evaporated under reduced pressure. The crude oily product (~ 62.4 g) is used for equilibration without further purification (Note 9).

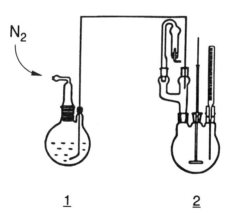

Figure I

b) Equilibration of ketones **2**. A solution of 62.4 g of crude **2** in 600 mL of ethanol, 80 mL of water and 70.0 g (1.2 mol) of potassium hydroxide is refluxed for 3 hr. The solution is concentrated on a rotary evaporator to a volume of about 200 mL and 500 mL of water is added. This aqueous solution is saturated with sodium chloride and extracted with four 100-mL portions of ether. The combined organic layers are dried over anhydrous magnesium sulfate and the solvent is evaporated at reduced pressure. The remaining oily liquid is distilled under reduced pressure at 0.05 mm. Three fractions are collected; the first fraction (boiling range 40-80°C) is discarded. Fraction 2 (boiling range: 80-100°C/120°C oil bath temperature) consists mainly of biphenyl with small amounts of ketone **2** (Note 10). Fraction 3 (boiling range: 100-110°C/140°C oil bath temperature) contains the main quantity of ketone **2**. Fraction 3 and the decanted liquid of fraction 2 are combined to yield 47.3-54.5 g (79-91%) of pale yellow oily **2** (Note 11).

B. *(1RS,2SR,5R)-5-Methyl-2-(1-methyl-1-phenylethyl)cyclohexanol* **3/4**. In a 500-mL, three-necked, round-bottomed reaction flask fitted with a reflux condenser carrying a calcium chloride tube, a 250-mL pressure-equalizing funnel and a mechanical Hershberg stirrer^2 are placed 16.0 g (0.70 mol) of sodium and 220 mL of toluene (Note 12). The solution is heated to reflux and maintained at this temperature. By vigorous stirring a fine suspension of sodium is obtained. To this stirred suspension a solution of 54.5 g (0.24 mol) of equilibrated **2** in 40.8 g (0.68 mol) of 2-propanol (Note 13) is added dropwise at such a rate that controlled refluxing is maintained. After the addition is complete the reaction mixture is refluxed for an additional 8 hr and then cooled to 0°C. The mixture is diluted with 250 mL of ether (Note 14) and carefully poured into 260 mL of ice-water. The organic layer is separated and the aqueous phase is saturated with sodium chloride and extracted three times with 100-mL portions of ether. The combined organic layers are washed with saturated aqueous sodium chloride solution, dried over anhydrous magnesium sulfate, filtered and concentrated by rotary evaporation. Fractional distillation of the concentrate gives 39.0-48.9 g (70-88%) of pale-yellow **3/4**, bp 103-107°C/0.01 mm (126°C oil bath temperature) (Note 15).

C. *(1R,2S,5R)-5-Methyl-2-(1-methyl-1-phenylethyl)cyclohexyl chloroacetate* **5** (Note 16). A 250-mL, three-necked, round-bottomed reaction flask fitted a with reflux condenser and calcium chloride tube, 50-mL pressure-equalizing funnel, thermometer and Teflon coated magnetic stirring bar is charged with 20.0 g (86 mmol) of **3/4**, 10.5 g (86 mmol) of N,N-dimethylaniline and 30 mL of diethyl ether. This stirred mixture is cooled to 0°C and a solution of 10.5 g (93 mmol) of chloroacetyl chloride in 30 mL of diethyl ether is added at such a rate that this temperature is maintained. After the reaction is stirred at 0°C for an additional 1 hr, the ice-bath is removed and the reaction mixture is allowed to warm to room temperature during which time N,N-dimethylaniline hydrochloride precipitates. The reaction is completed by heating to reflux for 3 hr (Note 17). The solvent is removed under reduced pressure using a rotary evaporator, and the crystalline white residue is dissolved in 60 mL of dichloromethane and 60 mL of water. The phases are separated and the organic phase is washed thoroughly with an equal volume of water; then it is washed until it is acid-free with a saturated aqueous sodium hydrogen carbonate solution. It is concentrated under reduced pressure to give about 25.0 g of a viscous oil, which crystallizes upon addition of 30 mL of 90% aqueous ethanol. The crystals are filtered with suction to yield 18.6-21.8 g (70-82%) of the chloroacetate as a mixture of diastereomers. Diastereo- and enantiomerically pure chloroacetate 5 is obtained in 48% yield by two fractional crystallizations of the diastereomeric chloroacetates from ethanol, mp 82-83°C; $[\alpha]_D^{20}$ +22.4° (CCl_4, *c* 2.29) (Note 18).

D. *(1R,2S,5R)-5-Methyl-2-(1-methyl-1-phenylethyl)cyclohexanol* **3**. In a 500-mL round-bottomed reaction flask, fitted with a reflux condenser and Teflon coated magnetic stirring bar, 12.8 g (41 mmol) of 5 (48%) is dissolved in a solution of 300 mL of ethanol, 40 mL of water and 4.6 g (82 mmol) of potassium hydroxide. This solution is refluxed for 2 hr. The solution is concentrated at reduced pressure to a volume of ca. 50 mL and 200 mL of water and 100 mL of ether are added. After the ether layer is separated, the aqueous phase is saturated with sodium chloride and extracted with three 50-mL portions of ether. The combined organic layers are dried over anhydrous magnesium sulfate, filtered, and the solvent is evaporated. Kugelrohr distillation of the cloudy residual oil yields 8.9-9.2 g (92-97%) of 3, bp 105-115°C/0.01 mm; $[\alpha]_D^{20}$ -26.4° ± 0.1° (ethanol, *c* 1.97) (Note 19).

2. Notes

1. Parts A and B of this procedure are based on a communication of E. J. Corey and H. E. Ensley.[3a]

2. The submitter used diethyl ether distilled from sodium wire.

3. Bromobenzene was purchased from Merck and Company, Inc. and was used without further purification.

4. Sometimes it becomes necessary to add some single crystals of iodine to start the reaction.

5. The concentration of the ethereal Grignard solution was estimated to be 1.38 N, as determined by hydrolysis of an aliquot (1 mL taken by syringe) and titration with 0.1 N hydrochloric acid.

6. The Teflon tube was 3 mm in diameter.

7. Copper(I) bromide was purchased from Fluka AG, Buchs, Switzerland and was not further purified. In previous runs copper(I) iodide was used to give comparable yields.

8. (R)-Pulegone had $[\alpha]_D^{20}$ +24.6° (ethanol, c 1.92) and was obtained from Haarmann & Reimer, Holzminden. The checkers used technical grade (+)-pulegone (82% pulegone content) and obtained 2 in 67-70% isolated yield after equilibration. The submitters thank Haarmann & Reimer, Holzminden, F. R. Germany, for generous gifts of pure and technical grade (+)-pulegone used in their work.

9. An epimeric mixture of diastereomeric ratio 55:45 was determined by ^{13}C NMR spectroscopy.

10. The Grignard-coupling product, biphenyl (mp 70°C, bp 250°C/760 mm), crystallizes in the condenser and has to be liquified by warming with a heat gun.

11. The elemental and structural characterization of 2 is as follows: Anal. Calcd for $C_{16}H_{22}O$: C, 83.43; H, 9.63. Found: C, 83.61; H, 9.80; IR (liquid film) cm^{-1}: 1712 (C=O); ^1H NMR (CDCl$_3$) δ: 0.80-2.78 (m, 8 H), 0.90 (d, 3 H, J = 6, C$\underline{H}_3$CH cis-2), 0.96 (d, 3 H, J = 6, C$\underline{H}_3$CH trans-2), 1.42 (s, 3 H, C$\underline{H}_3$CPh trans-2), 1.43 (s, 3 H, C$\underline{H}_3$CPh cis-2), 1.48 (s, 3 H, C$\underline{H}_3$CPh cis/trans-2), 7.00-7.44 (m, 5 H, aromatic H); ^{13}C NMR (CDCl$_3$), trans-2 δ: 22.21 (CH$_3$), 23.52 (CH$_3$), 26.71 (CH$_3$), 28.89 (CH$_2$), 34.51 (CH$_2$), 36.02 (CH), 38.93 (C$_{quat}$), 52.13 (CH$_2$), 59.23 (CH), 125.37, 125.60, 127.87 and 149.71 (aromatic C), 210.31 (C=O); cis-2 δ: 19.03 (CH$_3$), 23.67 (CH$_3$), 24.71 (CH$_2$), 27.23 (CH$_3$), 31.10 (CH$_2$), 32.00 (CH), 39.32 (C$_{quat}$), 50.07 (CH$_2$), 59.44 (CH), 125.44, 125.72, 128.65 and 149.26 (aromatic C), 211.21 (C=O). An epimeric mixture of diastereomeric ratio 83:17 was determined by ^{13}C NMR spectroscopy. Ketones 2 have n_D^{20} 1.5270-80.

12. Toluene was distilled from sodium.

13. 2-Propanol was refluxed with magnesium methoxide and fractionated.

14. Without this additional solvent the mixture is quite viscous.

15. ^{13}C NMR-spectroscopy indicated a 3/4-ratio of 87:13. Diastereomers 3/4 can be separated by careful medium-pressure silica gel chromatography (petroleum ether/ether; 95:5)[3b] or by fractional crystallization of the diastereomeric chloroacetates (vide supra). For structural characterization see Note 19.

16. This preparation has been modified.[3c] Compound 5 has also been prepared by using pyridine/4-dimethylaminopyridine in petroleum ether and chloroacetyl chloride in benzene.[4]

17. The submitters suggest that a revised work-up procedure as follows is preferable but it was not checked. At this point 250 mL of ether and 60 mL of water are added to dissolve the salt. The phases are separated and the procedure as described is followed.

18. The structural and elemental characterization of 5 is as follows: Anal. Calcd for $C_{18}H_{25}ClO_2$: C, 70.00; H, 8.16. Found: C, 69.81; H, 8.20; IR (KBr) cm^{-1}: 1185 (COC) and 1754 (C=O); ^{1}H NMR (CDCl$_3$) δ: 0.66-2.24 (m, 8 H), 0.90 (d, 3 H, J = 6, C$\underline{H}_3$CH), 1.22 (s, 3 H, C$\underline{H}_3$CPh), 1.33 (s, 3 H, C$\underline{H}_3$CPh), 3.04 and 3.43 (AB, 2 H, J = 15, C$\underline{H}_2$Cl), 4.91 (dt, 1 H, J = 10.6, 4, $\underline{H}$CO), 7.00-7.40 (m, 5 H, aromatic C$\underline{H}$); ^{13}C NMR (CDCl$_3$) δ: 21.71 (CH$_3$), 22.72 (CH$_3$), 26.18 (CH$_2$), 29.67 (CH$_3$), 31.22 (CH), 34.37 (CH$_2$), 39.36 (C$_{quat}$), 40.66 (CH$_2$), 41.43 (CH$_2$), 50.20 (CH), 75.65 (CH), 124.93, 125.09, 127.79 and 151.43 (aromatic C), 166.17 (C=O).

The submitters report that 2 may be recovered from the mother liquors enriched in the unwanted diastereomer. The mother liquors were saponified to 3/4 as described in section D, followed by dichromate oxidation to 2 in 77-81% yield according to a procedure given for menthone.[5] However, this recovery was not checked.

19. The optical rotation, $[\alpha]_D^{23}$ +26.3° (ethanol, c 2.02), for the enantiomer of 3 is reported.[3a] The elemental and structural characterization of 3 is as follows: Anal. Calcd for $C_{16}H_{24}O$: C, 82.70; H, 10.41. Found: C, 82.80; H, 10.27; IR (liquid film) cm^{-1}: 3420 and 3570 (OH); ^{1}H NMR (CDCl$_3$) δ: 0.64-2.06 (m, 9 H), 0.87 (d, 3 H, J = 6, C$\underline{H}_3$CH), 1.29 (s, 3 H, C$\underline{H}_3$CPh), 1.42 (s, 3 H, C$\underline{H}_3$CPh), 3.48 (dt, J = 10, 4, $\underline{H}$CO), 6.97-7.46 (m, 5 H, aromatic H); ^{13}C NMR (CDCl$_3$) δ: 21.95 (CH$_3$), 25.93 (CH$_3$), 26.62 (CH$_2$), 27.33 (CH$_3$), 31.45 (CH), 34.81 (CH$_2$), 39.87 (C$_{quat}$), 45.62 (CH$_2$), 53.85 (CH), 72.52 (CH), 125.22, 125.52, 127.88 and 150.83 (aromatic C).

3. Discussion

Since its introduction in 1975 by E. J. Corey and H. E. Ensley[3a] 8-phenylmenthol has found widespread use as a chiral auxiliary in organic syntheses.[3] It has proved to be dramatically superior in diastereoface discriminating ability to the commonly used chiral auxiliaries such as menthol, borneol, etc.

However, in spite of its well-documented applicability in asymmetric synthesis, no fully detailed paper has been published so far, concerning the preparation and isolation of enantio- and diastereomerically pure (-)-8-phenylmenthol.

Starting from (R)-pulegone we present herein an efficient three-step synthesis furnishing (-)-phenylmenthol as an easily separable 87:13-mixture of diastereomers in 55-80% overall isolated yield. Separation of the two diastereomers is achieved either by careful medium-pressure chromatography as Corey and Ensley stated,[3a,b] or, less tediously for greater quantities, by fractional crystallization of the chloroacetic acid esters and successive saponification as described herein.

Since the conversion of (-)-citronellol to (S)-pulegone is reported,[6] the enantiomeric (+)-8-phenylmenthol likewise may be synthesized. The latter should also be obtainable in a seven-step synthesis starting from (R)-pulegone (48% overall yield) as Corey and co-workers claimed.[3b]

1. Institut für Organische Chemie der Universität, Tammannstrasse 2, D-3400 Göttingen, Federal Republic of Germany.
2. Pinkney, P. S. *Org. Synth., Collect. Vol. II* **1943**, 116.
3. (a) Corey, E. J.; Ensley, H. E. *J. Am. Chem. Soc.* **1975**, *97*, 6908; (b) Ensley, H. E.; Parnell, C. A.; Corey, E. J. *J. Org. Chem.* **1978**, *43*, 1610; (c) Langström, B.; Stridsberg, B.; Bergson, G. *Chemica Scripta* **1978-1979**, *13*, 49; (d) Kaneko, T.; Turner, D. L.; Newcomb, M.; Bergbreiter, D. E. *Tetrahedron Lett.* **1979**, 103; (e) Oppolzer, W.; Robbiani, C.; Bättig, K. *Helv. Chim. Acta* **1980**, *63*, 2015; (f) Boeckman, Jr., R. K.; Naegely, P. C.; Arthur, S. D. *J. Org. Chem.* **1980**, *45*, 752; (g) Quinkert, G.; Schwartz, U.; Stark, H.; Weber, W.-D.; Baier, H.; Adam, F.; Dürner, G. *Angew. Chem.* **1980**, *92*, 1062; *Angew. Chem., Intern. Ed. Engl.* **1980**, *19*, 1029; (h) Oppolzer, W.; Kurth, M.; Reichlin, D.; Chapuis, C.; Mohnhaupt, M.; Moffatt, F. *Helv. Chim. Acta* **1981**, *64*, 2802; (i) Oppolzer, W.; Löher, H. J. *Helv. Chim. Acta* **1981**, *64*, 2808; (j) Whitesell, J. K.;

Bhattacharya, A.; Henke, K. *J. Chem. Soc., Chem. Commun.* **1982**, 988; (k) Whitesell, J. K.; Bhattacharya, A.; Aguilar, D. A.; Henke, K. *J. Chem. Soc., Chem. Commun.* **1982**, 989; (l) Roush, W. R.; Gillis, H. R.; Ko, A. I. *J. Am. Chem. Soc.* **1982**, *104*, 2269; (m) Quinkert, G.; Schwartz, U.; Stark, H.; Weber, W.-D.; Adam, F.; Baier, H.; Frank, G.; Dürner, G. *Liebigs Ann. Chem.* **1982**, 1999; (n) Taber, D. F.; Raman, K. *J. Am. Chem. Soc.* **1983**, *105*, 5935; (o) Whitesell, J. K.; Deyo, D.; Bhattacharya, A. *J. Chem. Soc., Chem. Commun.* **1983**, 802; (p) Koch, H.; Runsink, J.; Scharf, H.-D. *Tetrahedron Lett.* **1983**, *24*, 3217.

4. Ort, O. Diplomarbeit, Universität Göttingen, 1981.
5. Sandborn, L. T. *Org. Synth. Collect. Vol. I, 2nd ed.* **1941**, 340.
6. Corey, E. J.; Ensley, H. E.; Suggs, J. W. *J. Org. Chem.* **1976**, *41*, 380.

Appendix

Chemical Abstracts Nomenclature (Collective Index Number); (Registry Number)

(-)-8-Phenylmenthol: Cyclohexanol, 5-methyl-2-(1-methyl-1-phenylethyl)-, [1R-(1α,2β,5α)]- (10); (65253-04-5)

(2R,5R)-5-Methyl-2-(1-methyl-1-phenylethyl)cyclohexanone: Cyclohexanone, 5-methyl-2-(1-methyl-1-phenylethyl)-, (2R-trans)- (10); (57707-92-3)

(2S,5R)-5-methyl-2-(1-methyl-1-phenylethyl)cyclohexanone: Cyclohexanone, 5-methyl-2-(1-methyl-1-phenylethyl)-, (2S-cis)- (10); (65337-06-6)

Magnesium (8,9); (7439-95-4)

Bromobenzene: Benzene, bromo- (8,9); (108-86-1)

Copper(I) bromide: Copper bromide (8,9); (7787-70-4)

(R)-(+)-Pulegone: p-Menth-4(8)-en-3-one, (R)-(+)- (8); Cyclohexanone, 5-methyl-2-(1-methylethylidene)-, (R)- (9); (89-82-7)

Sodium (8,9); (7440-23-5)

2-Propanol (8,9); (67-63-0)

(1R,2S,5R)-5-Methyl-2-(1-methyl-1-phenylethyl)cyclohexyl chloroacetate: Acetic acid, chloro-, 5-methyl-2-(1-methyl-1-phenylethyl)cyclohexyl ester, [1R-(1α,2β,5α)]- (10); (71804-27-8)

N,N-Dimethylaniline: Aniline, N,N-dimethyl-(8); Benzenamine, N,N-dimethyl- (9): (121-69-7)

Chloroacetyl chloride: Acetyl chloride, chloro- (8,9); (79-04-9)

CHIRAL 1,3-OXATHIANE FROM (+)-PULEGONE: HEXAHYDRO-4,4,7-TRIMETHYL-4H-1,3-BENZOXATHIIN

(4H-1,3-Benzoxathiin, hexahydro-4,4,7-trimethyl-)

A. [structure] + $C_6H_5CH_2SH$ $\xrightarrow[\text{reflux}]{\text{NaOH, THF}}$ [structure]

B. [structure] $\xrightarrow[\substack{CH_3OH \\ -78°C}]{Na, NH_3}$ [structure]

C. [structure] $\xrightarrow[\substack{p\text{-}CH_3C_6H_4SO_3H \\ C_6H_6 \\ \text{reflux}}]{(CH_2O)_n}$ [structure]

Submitted by Ernest L. Eliel, Joseph E. Lynch, Fumitaka Kume, and Stephen V. Frye.[1]

Checked by Leticia M. Diaz, Stan S. Hall, and Gabriel Saucy.

1. Procedure

Caution! Benzyl mercaptan (Part A) is a foul-smelling liquid. Benzyl mercaptan, the liquid ammonia required in Part B and the benzene employed as a solvent in Part C should be used only in a well-ventilated hood.

A. *cis-* and *trans-5-Methyl-2-[1-methyl-1-(phenylmethylthio)ethyl]-cyclohexanone (7-benzylthiomenthone)*. A 1-L, three-necked, round-bottomed flask equipped with a magnetic stirring bar and a Friedrich condenser connected by its upper joint to a mineral oil bubbler through which passes nitrogen gas is charged with 500 mL of tetrahydrofuran (Note 1), 202.0 g (1.33 mol) of (+)-pulegone (Note 2) and 181.0 g (1.46 mol) of benzyl mercaptan (Note 3). The flask is flushed with nitrogen, 10 mL of aqueous 10% sodium hydroxide is added, the flask is stoppered and the mixture is heated to reflux under a static pressure of nitrogen. After 2 hr at reflux the pale yellow solution is allowed to cool, transferred to a 2-L separatory funnel and washed with two 500-mL portions of saturated aqueous sodium chloride. The combined sodium chloride layers, in turn, are extracted with three 250-mL portions of ether, and the combined organic layers are dried over magnesium sulfate for 2 hr, filtered and concentrated by rotary evaporation at aspirator vacuum. The residual liquid is distilled in a good vacuum to give a fraction of bp 167-174°C (0.4 mm) which weighs 325-330 g (89-90%). The product is a mixture of cis and trans isomers otherwise highly pure as evidenced by spectral analysis (Note 4).

B. *2-(1-Mercapto-1-methylethyl)-5-methylcyclohexanol (7-thiomenthol)*. A 5-L, three-necked, round-bottomed flask is equipped with a variable speed Hershberg stirrer and a 500-mL Dewar condenser filled with dry ice-acetone and connected by its upper joint to a mineral oil bubbler through which passes dry nitrogen gas. The flask is immersed in a dry ice-acetone bath, flushed well with nitrogen and 3000 mL of ammonia is condensed into the flask via a glass tube passed through a rubber septum in the remaining neck of the flask (Note 5). Clean sodium, 125 g (5.43 g-atom), is added slowly to the ammonia with slow stirring (Note 6). Then 250.3 g (0.906 mol) of *5-methyl-2-[1-methyl-1-(phenylmethylthio)ethyl]cyclohexanone and 72.5 mL (1.8 mol) of methanol in 625 mL of anhydrous ether (Note 7) are added dropwise via a pressure-equalized addition funnel over 5 hr to the vigorously stirred (ca. 500 rpm) solution (Note 8). Stirring is continued an additional 30 min following which 150 mL of methanol is added over 2.5 hr dropwise (to avoid a violent eruption). The solution is allowed to warm slowly (Note 9) and the addition funnel and the condenser are removed to allow the ammonia to evaporate overnight. The reaction flask is immersed in an ice bath and 700 mL of water is added cautiously over an hour to the yellow solid left by evaporation of the ammonia (Note 10). The solution is transferred to a 2-L separatory funnel and extracted with two 200-mL portions of ether, which are discarded. The aqueous layer is poured into a mixture of 500 mL of concentrated hydrochloric acid and 1000 g of ice, transferred to a 4-L separatory funnel and extracted with four 200-mL portions of ether. The combined ether extracts are washed with 200 mL of water, 200 mL of saturated aqueous sodium chloride, dried over magnesium sulfate and concentrated by rotary evaporation at aspirator vacuum. The residual liquid is placed under reduced pressure (0.2 mm) for 1 hr to remove the remaining solvent to give 137-140 g (80-82%) of an orange oil that is a diastereomeric mixture of which the major component constitutes 80%, as indicated by ^{13}C NMR (Note 11).

C. *Hexahydro-4,4,7-trimethyl-4H-1,3-benzoxathiin*. To a 1-L, one-necked, round-bottomed flask equipped with a magnetic stirrer and charged with 325 mL of benzene (Note 12) is added 140.0 g (0.753 mol) of *5-methyl-2-(1-methyl-1-thioethyl)cyclohexanol, 26.0 g (0.87 mol) of paraformaldehyde (Note 13) and 1 g of p-toluenesulfonic acid monohydrate (Note 14). The flask is fitted with a Dean-Stark trap and a Friedrich condenser and the contents are refluxed for

4 hr by which time the benzene distillate is clear. After the solution is cooled, 5 g of anhydrous potassium carbonate is added and the solution is stirred overnight, filtered, concentrated by rotary evaporation, and the residual liquid is distilled in a good vacuum to give a fraction of bp 69-94°C (0.1 mm), which weighs 130-134 g (86-89%) (Note 15). This fraction is dissolved in 250 mL of pentane, cooled to -25°C and seeded with a crystal of the product (Note 16). Crystallization is allowed to proceed in a freezer with collection and concentration of the mother liquor to half the original volume being carried out every other day to yield, after four crystallizations, 55-60 g (37-40%) of the spectrally pure product (Note 17).

2. Notes

1. Tetrahydrofuran was gold label from Aldrich Chemical Company, Inc.

2. (+)-Pulegone was obtained from SCM Corporation, Jacksonville, FL or Givaudan Corporation, Clifton, NJ; the specific rotation ranged from +21.85 to 22.6°. The material is also available from Aldrich Chemical Company, Inc. Pure pulegone has[2] $[\alpha]_D^{23}$ +22.5°. The discrepancy if any is probably due to chemical impurities since the pulegone used has been shown to be enantiomerically pure.[3,4] The checkers used (+)-pulegone from Givaudan Corporation, $[\alpha]_D^{20}$ +25.7°, which was 94.4% pure (GLC) and contained 4.8% of isopulegone or carvone.

3. Benzyl mercaptan was used as received from Aldrich Chemical Company, Inc.

4. Spectral data; IR (film), cm^{-1}: 1708, 1620, 1500, 1458, 1382, 1363, 1120, 710, and 695; ^{13}C NMR (50 MHz, CDCl$_3$) δ: 22.2, 23.8, 27.8, 29.6, 33.1, 34.5, 36.6, 48.0, 52.3, 57.8, 126.8, 128.4 (2 C), 128.9 (2 C), 138.7, and 210.2; ^{1}H NMR (200 MHz, CDCl$_3$) δ, partial: singlets at 1.38, 1.60, and 3.73, and an intense doublet at 0.97 (J = 5.9). The checkers noted that the product distillate was a pale blue color, which turned pale yellow after a few minutes.

5. The submitters noted that at least 15 kg of dry ice was required for a reaction of this scale. Using the described apparatus the checkers found that the distillation of the ammonia required 8-10 hr, which could be reduced to ca. 5 hr by using the following assembly. An oven-dried, 5-L, three-necked, round-bottomed flask was equipped with a glass stirrer shaft fitted with a sleeve joint and a large Teflon blade and the shaft was connected to an overhead motor drive. The flask was also connected to two 450-mL Dewar condensers with a large soda lime drying tube attached to the tube connector of one condenser while the tube connector of the other condenser was attached with Tygon tubing that led through a tower of solid potassium hydroxide pellets to a tank of anhydrous ammonia. While ammonia was slowly flushed through the entire assembly, the Dewar condensers were filled with dry ice-acetone and a dry ice-acetone bath was raised to cool the vessel by immersion. The ammonia flow-rate was increased to condense ca. 3000 mL of ammonia into the flask. Then the condenser with the inlet connection from the ammonia was removed and the flask sealed with a glass stopper.

6. Sodium was stored under mineral oil and washed with pentane before use. For convenience the checkers used 1/6 to 1/4 inch sodium spheres (Matheson Coleman and Bell) that were weighed in mineral oil, then wiped free of oil, rinsed in hexane, cut in half, rinsed in hexane again, and immediately added to the reaction over a 2-hr period, during which time the dark-black mixture became extremely viscous.

7. At this point the checkers charged an oven-dried, 1000-mL pressure-equalizing addition funnel with the ketone in methanol and ether, and then quickly mounted the sealed funnel on the reaction flask by removing the flask's glass stopper.

8. Methanol was used as received from Fisher Scientific, Inc. Anhydrous ether was used as received from freshly opened containers from Mallinckrodt, Inc. and Fisher Scientific, Inc.

9. It is important to let the reaction mixture warm slowly; otherwise the ammonia will boil violently and carry some of the reaction material out of the flask.

10. This reaction is highly exothermic and caution should be exercised since some active sodium may occasionally be left on the sides of the flask.

11. The intense ^{13}C NMR (50 MHz, $CDCl_3$) signals of the major isomer are at 21.9, 26.9, 29.0, 31.3, 34.5, 34.6, 45.4, 54.6 and 72.9 ppm relative to TMS; ^{1}H NMR (200 MHz, $CDCl_3$) δ, partial: 0.91 (d, 3 H, J = 6.5), 1.40 (s, 3 H), and 1.52 (s, 3 H).

12. Benzene was used as received from Aldrich Chemical Company, Inc. *Caution!* See benzene warning in *Org. Synth.* **1978**, *58*, 168.

13. Paraformaldehyde was used as received from Aldrich Chemical Company, Inc.

14. p-Toluenesulfonic acid monohydrate was used as received from Aldrich Chemical Company, Inc.

15. Excess paraformaldehyde may separate from the distillate. If this occurs, the liquid should be filtered prior to crystallization.

16. In the absence of seeding, crystallization may take several weeks. It is preferable to separate a small sample of the precursor thiomenthol from its stereoisomers by HPLC (3% ethyl acetate in hexane as eluant) and prepare a small amount of pure oxathiane from this material. Alternatively, a small amount of the product may be purified by GLC on a 5% FFAP column. The melting point of pure material is 37-38°C. The checkers, who did not have seeding crystals, found that the early crops of crystals melted when the flask was allowed to warm to ambient temperature. Consequently the cold supernatant liquid was withdrawn from the crystals with a Pasteur pipette while the flask was maintained at ca. 0°C (ice-water bath). The crystals were subsequently recrystallized several times in the same flask without filtration. By this technique, white crystals melting at 32-35°C were obtained; this material is spectrally pure and suitable for asymmetric synthesis. The supernatant liquid was also concentrated, as the submitters described, to obtain additional crops using this technique.

17. Spectral data; IR (film), cm^{-1}: 2970-2870, 1455, 1440, 1388, 1370, 1355, 1305, 1155, 1095, 1066, 985, 955, 900, 830, and 710; ^{13}C NMR (50 MHz, $CDCl_3$) δ: 21.8, 22.1, 24.4, 29.4, 31.3, 34.7, 41.8 (2 C), 51.5, 67.1, and 76.7 ppm; ^1H NMR (200 MHz, $CDCl_3$) δ partial: 0.92 (d, 3 H, J = 6.5), 1.27 (s, 3 H), 1.43 (s, 3 H), 3.35 (td, 1 H, J = 10.5, 4.2, HCO), 4.69 (d, 1 H, J = 11.5, SCH$_2$O), 5.03 (d, 1 H, J = 11.5, SCH$_2$O).

3. Discussion

Hexahydro-4,4,7-trimethyl-4H-benzoxathiin is used as a chiral template in the asymmetric synthesis, in over 90% enantiomeric excess, of tertiary[3,5] and secondary[6] α-hydroxy aldehydes, RR'C(OH)CHO and the derived acids, RR'C(OH)CO$_2$H and glycols, RR'C(OH)CH$_2$OH.[7] The present procedure is a slight modification of a published[5] one.

1. W. R. Kenan, Jr. Laboratories, Department of Chemistry, University of North Carolina, Chapel Hill, NC 27514.
2. Corey, E. J.; Ensley, H. E.; Suggs, J. W. *J. Org. Chem.* **1976**, *41*, 380.
3. Lynch, J. E.; Eliel, E. L. *J. Am. Chem. Soc.* **1984**, *106*, 2943-2948.
4. Whitesell, J. K., personal communication.
5. Eliel, E. L.; Lynch, J. E. *Tetrahedron Lett.* **1981**, *22*, 2855.
6. Ko, K.-Y.; Frazee, W. J.; Eliel, E. L. *Tetrahedron* **1984**, *40*, 1333-1343.
7. Eliel, E. L.; Koskimies, J. K.; Lohri, B.; Frazee, W. J.; Morris-Natschke, S.; Lynch, J. E.; Soai, K. In "Asymmetric Reactions and Processes in Chemistry", Am. Chem. Symposium Series #185; Eliel, E. L.; Otsuka, S., Eds.; Washington, 1981; p. 37; Eliel, E. L. *Phosphorus and Sulfur* **1985**, *24*, 73.

Appendix

Chemical Abstracts Nomenclature (Collective Index Number);

(Registry Number)

(+)-Pulegone: p-Menth-4-(8)-en-3-one, (R)-(+)- (8); Cyclohexanone,
5-methyl-2-(1-methylethylidene)-, (R)- (9); (89-82-7)

Hexahydro-4,4,7-trimethyl-4H-1,3-benzoxathiin: 4H-1,3-Benzoxathiin,
hexahydro-4,4,7-trimethyl- (9); (59324-06-0)

Benzyl mercaptan: α-Toluenethiol (8); Benzenemethanethiol (9); (100-53-8)

cis- and trans-5-Methyl-2-[1-methyl-1-(phenylmethylthio)ethyl]cyclohexanone:
Cyclohexanone, 5-methyl-2-[1-methyl-1-[(phenylmethyl]thio]ethyl]-,
(2R-trans)-(10); (79563-58-9); (2S-cis)- (10); (79618-04-5)

2-(1-Mercapto-1-methylethyl)-5-methylcyclohexanol: Cyclohexanol-2-
(1-mercapto-1-methylethyl)-5-methyl-, [1R-(1α, 2α, 5α)]- (10);
(79563-68-1); [1R-[1α, 2β, 5α)]- (10); (79563-59-0); [1S-(1α, 2α, 5β)]-
(10); (79563-67-0)

Ammonia (8,9); (7664-41-7)

Sodium (8,9); (7440-23-5)

Paraformaldehdye: Poly(oxymethylene) (8,9); (9002-81-7)

p-Toluenesulfonic acid monohydrate (8); Benzenesulfonic acid, 4-methyl-,
monohydrate (9); (6192-52-5)

(−)-α-PINENE BY ISOMERIZATION OF (−)-β-PINENE

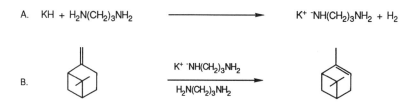

A. KH + $H_2N(CH_2)_3NH_2$ ⟶ $K^+\ ^-NH(CH_2)_3NH_2 + H_2$

B. [β-pinene] $\xrightarrow[H_2N(CH_2)_3NH_2]{K^+\ ^-NH(CH_2)_3NH_2}$ [α-pinene]

Submitted by Charles A. Brown[1] and Prabhakav K. Jadhav.[2]
Checked by Janet W. Grissom and Edwin Vedejs.

1. Procedure

Caution! Potassium hydride[3] is highly reactive toward water. Material separated from protective oil or solvent should not be exposed to air, but should be kept under argon or nitrogen.

A. *Potassium 3-aminopropylamide (KAPA).* To a bottle of 22.4% potassium hydride in oil (Note 1) is added a Teflon-covered magnetic stirring bar. The contents are agitated by first shaking and then stirring until a visually uniform dispersion is attained. Then 17.8 g (0.10 mol) of the dispersion is transferred (Note 2) to an oven-dried 250-mL flask, fitted with a magnetic stirring bar (Notes 3 and 4), a septum-capped inlet, and an adapter connected to a pressure relief bubbler (mercury or oil). The apparatus is purged with dry argon or nitrogen (Note 5). The flask is charged with 100 mL of dry pentane or other volatile hydrocarbon solvent, and the contents are stirred thoroughly. The stirrer is stopped to allow the potassium hydride to settle, assisted by gentle tapping on the flask, and the supernatant liquid is drawn off with a double-ended needle under inert gas pressure. This procedure is repeated twice to complete removal of the oil. Residual solvent is then removed in a stream of dry gas. To the dry potassium hydride powder is added rapidly 100 mL of dry 3-aminopropylamine (Note 6). *Caution! Hydrogen gas evolution.* Hydrogen evolution commences immediately (Note 7) and subsides after 1.5 to 2 hr. Formation of KAPA is complete at this time.

B. *(-)-α-Pinene.* Concurrently, a 3-L, three-necked, round-bottomed flask (fitted with a septum-capped adapter, gas-tight mechanical stirrer, and a tube adapter connected to a gas bubbler) is purged with inert gas (Notes 3 and 4). The flask is then charged with 1,588 mL of (-)-β-pinene (Note 8). To the vigorously stirred pinene at 25°C is added the KAPA (0.1 mol) prepared in Part A, using a double-ended needle, over 15 min. The reaction mixture is stirred for 24 hr at 25°C and is then quenched by addition of 100 mL of ice-cold water. The reaction mixture is transferred to a separatory funnel, washed with two 100-mL portions of water, and dried with anhydrous calcium chloride. The crude pinene is decanted from the calcium chloride through a plug of glass wool and distilled at reduced pressure from lithium aluminum hydride (Notes 8 and 9). The yield of (-)-α-pinene, bp 72-73°C (46 mm), $[\alpha]_D^{23}$ -47.45° (neat, 92% ee), is 1,264 g (93%). The chemical and isomeric purity is 99% by GLC (Notes 10 and 11).

2. Notes

1. Potassium hydride in oil is available from Aldrich Chemical Company, Inc. and Alfa Products, Morton/Thiokol Inc. Material from both sources has been used successfully in this preparation.

2. Potassium hydride dispersion is transferred with a 20- or 30-mL syringe fitted with a 16-gauge needle. The plunger should be lubricated with mineral oil. The density of the potassium hydride dispersion is nearly 1.0 g/mL. Alternatively, a glove box or bag may be used for direct weighing.

3. Teflon is attacked superficially by KAPA with darkening of the surface and of the solution in contact with it. The reactivity of the KAPA is not affected. Polyethylene is not affected at all.

4. All equipment should be dried in an oven and cooled in a desiccator or under dry nitrogen.

5. From this point until the quenching of the final reaction operations must be carried out under argon or nitrogen.

6. For this reaction the amine (Aldrich Chemical Company, Inc.) from a freshly opened bottle is dried by distillation from active powdered calcium hydride. Addition over 1 min or less is required to avoid excessive foaming which has been observed when slow or dropwise addition is used. Older samples of amine which contain water should be pre-dried overnight with potassium hydroxide.

7. A cool (10-15°C) bath should be available to moderate the reaction if foaming occurs. Some samples of potassium hydride which have been exposed to air react sluggishly. Here application of a warm (35-40°C) bath is helpful.

8. The (-)-β-pinene used in this procedure was obtained from Glidden Organics. It was purified before use by distillation from lithium aluminum hydride; small increments of hydride were added to pinene stirred under nitrogen until an excess was present as determined by observing gas evolution on addition of a few drops of methanol to an aliquot of the slurry. The purified material had $[\alpha]_D^{23}$ -21.0° (neat, 92.1% ee). GLC analysis (10% SE-30 on Chromosorb W, 100°C) showed only one peak. Use of dry, peroxide-free

pinene is critical to the success of this procedure. The (-)-β-pinene is also available from Aldrich Chemical Company, Inc. In separate experiments, (-)-β-pinene samples from Aldrich Chemical Company, Inc. and Fluka A.G. were used successfully. The checkers used Aldrich β-pinene and obtained product with $[\alpha]_D^{30}$ - 46.3°.

9. Distillation from lithium aluminum hydride serves to remove completely all traces of water and amine as is required for preparation of chiral organoboranes.

10. For GLC conditions see Note 8.

11. The reaction has been carried out at 0°C on a 0.33-mol scale using a vigorous magnetic stirrer for 4 hr.[4] Under these conditions the isomeric purity was >99.5%. Because the reaction takes place in a two-phase system, the reaction rate appears to be stirrer-dependent and vigorous agitation is essential.

3. Discussion

β-Pinene is an important auxiliary for directed chiral syntheses. It has been used for preparation of mono- and diisopinocampheylborane[5] B-allyldiisopinocampheylborane,[6] B-pinanyl-9-bora[bicyclo]nonane,[5] cis-pinanediol,[7] and 2-hydroxypinan-3-one.[8]

Although both enantiomers of α-pinene occur in nature, only the (+)-enantiomer is commercially available with acceptable enantiomeric purity (92-95%). The (-)-enantiomer is only readily available with 80-85% enantiomeric purity, which is impractical for use in asymmetric syntheses via organoboranes.

Fortunately, (-)-β-pinene of high enantiomeric purity (90-95%) is commercially available and can be isomerized to the thermodynamically more stable (-)-α-pinene. The isomerization has been achieved by use of several acidic, basic, and metal catalysts.[4] These methods have various limitations such as partial racemization, low chemical yields, and vigorous reaction conditions. The current procedure is based on our earlier report using KAPA at 0°C on a much smaller scale.[4] The change to 25°C from 0°C was made to simplify the reaction conditions by eliminating the need for external cooling. The reaction has also been carried out on somewhat reduced scale (1-5 mol) using a magnetic stirrer in place of the mechanical stirrer.

1. Chemical Dynamics Dept., IBM Research Laboratory K34-281, 5600 Cottle Road, San Jose, CA 95193.
2. Richard B. Wetherill Chemistry Laboratory, Purdue University, West Lafayette, IN 47907.
3. For a general discussion on metallation with and handling of potassium hydride see: Brown, C. A. *J. Org. Chem.* **1974**, *39*, 3913.
4. For a general discussion on methods of isomerization of β-pinene to α-pinene see: Brown, C. A. *Synthesis* **1978**, 754.
5. Brown, H. C.; Jadhav, P. K.; Mandel, A. K. *Tetrahedron* **1981**, *37*, 3547.
6. Brown, H. C.; Jadhav, P. K. *J. Am. Chem. Soc.* **1983**, *105*, 2092.
7. Matteson, D. S.; Ray, R. *J. Am. Chem. Soc.* **1980**, *102*, 7590.
8. Yamada, S.-i.; Oguri, T.; Shioiri, T. *J. Chem. Soc., Chem. Commun.* **1976**, 136.

Appendix

Chemical Abstracts Nomenclature (Collective Index Number);

(Registry Number)

(-)-α-Pinene: 2-Pinene, (1S,5S)-(-) (8); Bicyclo[3.1.1]hept-2-ene, 2,6,6-trimethyl-, (1S)- (9); (7785-26-4)

(-)-β-Pinene: Bicyclo[3.1.1]heptane, 6,6-dimethyl-2-methylene-, (1S)- (9); (18172-67-3)

Potassium hydride (8,9); (7693-26-7)

3-Aminopropylamine: 1,3-Propanediamine (8,9); (109-76-2)

DIISOPROPYL (2S,3S)-2,3-O-ISOPROPYLIDENETARTRATE

(1,3-Dioxolane-4,5-dicarboxylic acid, 2,2-dimethyl-, bis(1-methylethyl)ester, (4R-trans)-)

Submitted by René Imwinkelried, Martin Schiess, and Dieter Seebach.[1]
Checked by Isao Kurimoto and Ryoji Noyori.

1. Procedure

A dry, 500-mL, two-necked flask equipped with a magnetic stirrer and a reflux condenser is flushed with nitrogen and charged with 8.4 mL (45.8 mmol) of dimethyl (2S,3S)-2,3-O-isopropylidenetartrate (Note 1) and 250 mL of absolute 2-propanol (Note 2). To the resulting solution is added with a plastic syringe and hypodermic needle 1.35 mL (4.6 mmol) of tetraisopropyl titanate (Note 1). The mixture is refluxed with stirring for 2 hr. To remove the methanol formed, the flask is transferred to a rotary evaporator, and the contents are concentrated to 10 to 12 mL. The oily residue is once more dissolved in 250 mL of absolute 2-propanol (Note 2) and refluxed for 2 hr. The solvent is removed again in a rotary evaporator, and the resulting yellow oil is dissolved in 100 mL of diethyl ether. After addition of 5 mL of water (Note 3) the pale mixture is vigorously stirred for 10 min and then dried over anhydrous magnesium sulfate. The flaky suspension is filtered and the filter-cake washed with three 25-mL portions of ether. The ether solution is concentrated in a rotary evaporator. The residue, 12.5-13.1 g of a slightly yellow oil, solidifies on standing. This solid is freed from small amounts of solvent by an oil-pump vacuum (ca. 0.01 mm) at room temperature for 2 hr. Further purification by short-path distillation at 91-93°C/0.05 mm furnishes 11.5-12.0 g (91-95%) of a slightly yellow solid, which turns colorless upon crushing, mp 41.5-42.5°C, $[\alpha]_D^{rt}$ +42 ± 0.3° (CHCl$_3$, c 4).

2. Notes

1. Commercial (Fluka purum) (-)- or (+)-dimethyl 2,3-O-isopropylidene-tartrate and tetraisopropyl titanate can be used without further purification.

2. 2-Propanol was heated at reflux over CaSO$_4$, distilled, and redistilled with addition of tetraisopropyl titanate (ca. 10 g/L).

3. This is done to hydrolyze titanium alkoxides. Part of the titanium alkoxides is removed during evaporation of the solvents in the rotatory evaporator [Ti(OCHMe$_2$)$_4$, bp 78°C/12 mm].

3. Discussion

Normally, transesterifications are acid- or base-catalyzed, e.g., sulfuric acid, p-toluenesulfonic acid, and potassium or sodium alkoxides in the appropriate alcohols.[2] These methods fail with molecules containing acid- or base-labile functional groups. The titanate-mediated esterifications, deacylations, and transesterifications of rather simple, monofunctional substrates are described in the patent literature; see the references in a recent article.[3] Recently, Seebach, et al.[3,4,5] have demonstrated that this method is applicable also to substrates with additional functional groups which would not survive acid- or base-catalyzed transesterification conditions, such as $C\equiv C$ and $C\equiv N$ bonds, acetals, β-hydroxy and β-acyloxy esters, β-lactams, tert-butyldimethylsilyloxy groups, BOC[6] and other carbamate protecting groups, etc. In the accompanying Scheme, the possible applications of this transesterification are illustrated, and some characteristic examples are given in the Table. Of course, the method can only establish equilibrium conditions. Therefore, depending on the particular case, components of the equilibria have to be removed (see procedure above) or used in large excess to drive the conversion to the desired products.

Scheme

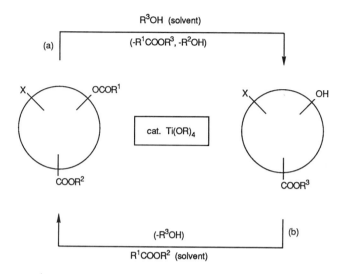

Titanate-mediated Transesterifications

X = Functional group (see accompanying text)

(a) Transesterification in alcoholic solvents, with removal of acyl protecting groups and exchange of the alcohol component of ester groups in the substrate.

(b) Transesterification in ester solvents, with acylation of hydroxy groups and exchange of the alcohol or of the acid component of ester groups in the substrate.

Table I. Products of Transesterification with Titanate Catalysis[3-6]

71% from ethyl ester
and 2-trimethylsilylethanol

74% from methyl ester
and 2-trimethylsilylethanol

91% from methyl ester
and 2-propanol

50% from methyl ester
and 2-propanol

90% from 0-3,5-dinitrobenzoate
and ethanol

60% from 0-propanoyl
derivative and ethanol

88% from ethyl ester
and methanol

70% from alcohol
and ethyl acetate

83% from ethyl ester
and 2-propanol

89% from methyl ester
and benzyl alcohol

1. Laboratorium für Organische Chemie der Eidgenössischen Technischen Hochschule, ETH-Zentrum, Universitätstrasse 16, CH-8092 Zürich, Switzerland.
2. Patai, S., Ed. "The Chemistry of Acid Derivatives", Supplement B, Part 1; Interscience Publishers: New York, 1979.
3. Seebach, D.; Hungerbühler, E., Naef, R.; Schnurrenberger, P.; Weidmann, B.; Züger, M. F. *Synthesis* **1982**, 138.
4. Schnurrenberger, P.; Züger, M. F.; Seebach, D. *Helv. Chim. Acta* **1982**, *65*, 1197. See also: Seebach, D.; Züger, M. *Helv. Chim. Acta* **1982**, *65*, 495.
5. Seebach, D.; Weidmann, B.; Widler, L. In "Modern Synthetic Methods 1983", Scheffold, R., Ed.; Otto Salle: Frankfurt, Sauerländer: Aarau, Wiley: New York, 1983; Vol. 3, p. 217.
6. Rehwinkel, H.; Steglich, W. *Synthesis* **1982**, 826.

Appendix
Chemical Abstracts Nomenclature (Collective Index Number);
(Registry Number)

Diisopropyl (2S,3S)-2,3-O-isopropylidenetartrate: 1,3-Dioxolane-4,5-dicarboxylic acid, 2,2-dimethyl-, bis(1-methylethyl) ester, (4R-trans)- (11); (81327-47-1)

Dimethyl (2S,3S)-2,3-O-isopropylidenetartrate: 1,3-Dioxolane-4,5-dicarboxylic acid, 2,2-dimethyl-, dimethyl ester, (4R-trans)- or (4S-trans)- (9); (37031-29-1) or (37031-30-4)

2-Propanol (8,9); (67-63-0)

Tetraisopropyl titanate: Isopropyl alcohol, titanium (4+) salt (8); 2-Propanol, titanium (4+) salt (9); (546-68-9)

1-DEOXY-2,3,4,6-TETRA-O-ACETYL-1-(2-CYANOETHYL)-α-D-GLUCOPYRANOSE

(D-glycero-D-ido-Nonononitrile, 4,8-anhydro-2,3-dideoxy-, 5,6,7,9-tetraacetate)

Submitted by Bernd Giese, J. Dupuis, and Marianne Nix.[1]
Checked by John P. Daub and Bruce E. Smart.

1. Procedure

Caution! Acrylonitrile is an OSHA-regulated carcinogen and appropriate precautions should be taken in handling this substance.

A 250 mL, three-necked, round-bottomed flask equipped with a magnetic stirring bar, dry nitrogen inlet, reflux condenser, and septum is flushed with nitrogen and charged with 20.6 g (50 mmol) of 2,3,4,6-tetra-O-acetyl-α-D-glucopyranosyl bromide (Note 1) and 100 mL anhydrous ether. The mixture is brought to reflux with a heat gun. A nitrogen atmosphere is maintained over the reaction mixture during this and the ensuing steps. When the 2,3,4,6-tetra-O-acetyl-α-D-glucopyranosyl bromide dissolves, 13.5 g (250 mmol) of acrylonitrile and 16.0 g (55 mmol) of tributyltin hydride (Note 2) are added by means of a syringe. The refluxing solution is irradiated with a sunlamp positioned 2.5-3.8 cm from the reaction flask (Note 3). The heat from the sunlamp maintains a vigorous reflux. After 4 hr, the irradiation is stopped and the reaction mixture is cooled and filtered to collect the precipitated crystals (Note 4). The filtrate is reintroduced under nitrogen into the reaction vessel. An additional 6.6 g (120 mmol) of acrylonitrile and 5.8 g (20 mmol) of tributyltin hydride are added and irradiation is resumed for another 4 hr until the starting bromide is completely reacted (Note 5). After the reaction mixture is cooled, it is filtered, the filtrate is saved, and the filter cake (Note 6) is combined with the crystals collected above. The solids are stirred with 250 mL of hot isopropyl alcohol and filtered. The hot filtrate is concentrated to a total volume of 75 mL, allowed to cool and filtered to give 7.8-8.6 g (40-45%) of 1-deoxy-2,3,4,6-tetra-O-acetyl-1-(2-cyanoethyl)-α-D-glucopyranose as long, colorless crystals, mp 121-122°C, $[\alpha]_D^{25}$ +66.2 ($CHCl_3$, c 0.7) (Note 7). The isopropyl alcohol mother liquor is set aside. The above filtrate from the crude reaction mixture is concentrated, taken up in 50 mL of acetonitrile, and extracted three times with 50-mL portions of pentane. The acetonitrile phase is combined with the isopropyl alcohol mother liquor and concentrated. The resulting syrup is flash-chromatographed on silica gel using ethyl acetate/hexane (1:1) eluent to afford an additional 2.2-2.5 g of pure 1-deoxy-2,3,4,6-tetra-O-acetyl-1-(2-cyanoethyl)-α-D-glucopyranose after separation of byproducts (Notes 8 and 9). The combined yield of pure product is 53-56%.

2. Notes

1. This material was obtained from the Sigma Chemical Company and was recrystallized from ethyl ether/pentane before use. It also can be prepared by the procedure of Redemann, C. E.; Niemann, C. *Org. Synth.*, *Collect. Vol. III* **1955**, 11.

2. Acrylonitrile and tributyltin hydride were obtained from the Aldrich Chemical Company, Inc., and used without further purification. The use of a large excess of acrylonitrile reduced the amount of reduction by-product (2,3,4,6-tetraacetyl-1,5-anhydroglucitol). Because tributyltin hydride also reacts with acrylonitrile, a small excess must be used.

3. The checkers used a 275-watt General Electric sunlamp. The submitters indicate that the type of sunlamp is not critical and that a 125-watt Philips E99/2 sunlamp or a 125-watt Philips 57203 high-pressure mercury lamp are suitable. If the heat from the sunlamp does not maintain reflux, the reaction flask should be about one-fourth submerged into a hot oil bath.

4. During the reaction the amount of precipitate gradually increases until the light from the sunlamp is totally blocked and the reaction stops.

5. The reaction can be followed by TLC using 0.25-mm silica gel 60 F-254 plates (E. Merck & Company) and ethyl acetate/hexane (1:1) eluent: 2,3,4,6-tetra-O-acetyl-α-D-glucopyranosyl bromide (R_f = 0.39); 2,3,4,6-tetra-O-acetyl-1,5-anhydroglucitol (R_f = 0.28); 1-deoxy-2,3,4,6-tetra-O-acetyl-1-(2-cyanoethyl)-β-D-glucopyranose (R_f = 0.20); 1-deoxy-2,3,4,6-tetra-O-acetyl-1-(2-cyanoethyl)-α-D-glucopyranose (R_f = 0.16).

6. This solid contains a large amount of insoluble polymer in addition to product.

7. The product is analytically pure: Anal. Calcd for $C_{17}H_{23}NO_9$: C, 52.98; H, 6.02; N, 3.63. Found: C, 52.80; H, 6.01; N, 3.63; ^{1}H NMR (CDCl$_3$, 360 MHz) : 1.84-1.95 (m, 1 H, CC$\underline{H}_2$), 2.05 (s, 6 H, 2 OAc), 2.09 (s, 3 H, OAc), 2.11 (s, 3 H, OAc), 2.05-2.22 (m, 1 H, CC$\underline{H}_2$), 2.46 (m, 2 H, C$\underline{H}_2$CN), 3.88 (ddd, 1 H, $\underline{H}_5$, $J_{5,6}$ = 5.80, $J_{5,6'}$ = 2.88, $J_{4,5}$ = 8.30), 4.12 (dd, 1 H, $\underline{H}_{6'}$, $J_{6,6'}$ = 12.24), 4.23 (m, 1 H, $\underline{H}_1$), 4.32 (dd, 1 H, $\underline{H}_6$), 4.98 (t, 1 H, $\underline{H}_4$, $J_{3,4}$ = 8.30), 5.09 (dd, 1 H, $\underline{H}_2$, $J_{1,2}$ = 5.23, $J_{2,3}$ = 8.30), 5.23 (t, 1 H, $\underline{H}_3$); IR (KBr) cm^{-1}: 2240 (C N), 1745 (C=O).

8. E. Merck 230-400 mesh Kieselgel 60 silica gel was used in the flash-chromatography. The first fraction contained 3.5 g (21%) of the tetra-O-acetyl-1,5-anhydroglucitol by-product; the second fraction contained 0.95-1.00 g (5%) of pure 1-deoxy-2,3,4,6-tetra-O-acetyl-1-(2-cyanoethy)-β-D-glucopyranose: mp 91-93°C, 117-119°C (two forms); $[\alpha]_D^{25}$ -19.6 (CHCl$_3$, c 1.01). The final fraction contained the desired product. The ratio of α:β isomers based on isolated yields is 11:1. The submitters report a ratio of 14:1 based on chromatographic analysis of the crude reaction mixture.

9. 1-Deoxy-2,3,4,6-tetra-O-acetyl-1-(2-cyanoethyl)-β-D-glucopyranose shows IR (KBr) cm^{-1}: 2240 (C≡N), 1745 (C=O); ^{1}H NMR (CDCl$_3$, 360 MHz) δ: 1.79 (br m, 1 H, CCH$_2$), 1.92 (br m, 1 H, CCH$_2$), 2.00 (s, 3 H, OAc), 2.03 (s, 3 H, OAc), 2.07 (s, 3 H, OAc), 2.09 (s, 3 H, OAc), 2.52 (m, 2 H, CH$_2$CN), 3.56 (dt, 1 H, H$_1$, J$_{1,2}$ = 9.6, J = 9.6, 2.6), 3.68 (ddd, 1 H, H$_5$, J$_{4,5}$ = 9.9, J$_{5,6}$ = 5.0, J$_{5,6'}$ = 2.1), 4.11 (dd, 1 H, H$_{6'}$, J$_{6,6'}$ = 12.2), 4.24 (dd, 1 H, H$_6$), 4.88 (t, 1 H, H$_2$, J$_{2,3}$ = 9.6), 5.05 (t, 1 H, H$_4$, J$_{3,4}$ = 9.4), 5.20 (t, 1 H, H$_3$). Anal. Calcd for C$_{17}$H$_{23}$NO$_9$: C, 52.98; H, 6.02; N, 3.63. Found: C, 52.85; H, 6.09; N, 3.70.

3. Discussion

The formation of the C-C bond occurs in a radical chain reaction[2] (Scheme 1). Bromine abstraction from halide **1** by tin radicals **2**[3] leads to carbon radicals **3** that react with alkenes **4** to give product radicals **5**. Trapping of **5** by tributyltin hydride yields products **6** and the tin radical **2**.

Scheme 1

$$Bu_3SnH \longrightarrow Bu_3Sn\cdot$$
$$ 2$$

$$Bu_3Sn\cdot + RX \longrightarrow R\cdot + Bu_3SnX$$
$$ 1 3$$

$$R\cdot + H_2C=CHCN \longrightarrow RCH_2\dot{C}HCN$$
$$3 4 5$$

$$RCH_2\dot{C}HCN + Bu_3SnH \longrightarrow RCH_2CH_2CN + Bu_3Sn\cdot$$
$$5 6$$

In contrast to anionic 1,4-addition methods, the radical procedure tolerates an acetoxy group adjacent to the reactive center (radical - bearing carbon). In applying this general method to tetraacetylglucosyl bromide 7,[4] products 9, with an axial C-C bond, and 10 are obtained in a ratio of ca. 10:1. ESR experiments show that the radical intermediate has the boat conformation 8.[5] The attack that leads to the axial product 9 is, therefore, the favored pesudoequatorial attack,[6] assuming that the transition state also resembles boatlike 8. Some of the radical 8 is trapped directly by tributyltin hydride to give the simple reduction product 11. This is a common side-reaction in tin hydride-initiated radical coupling reactions.

240

Scheme 2

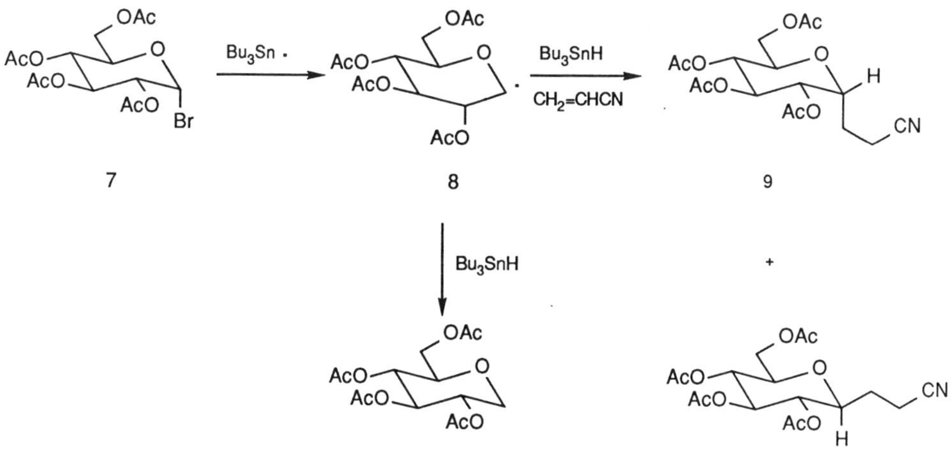

This method has been applied also to mannosyl bromide and galactosyl bromide.[4] Because alkoxyalkyl radicals are nucleophilic radicals, only alkenes with electron-withdrawing substituents can be used.[7] The 1,5-anhydroglycitol side product **11** is formed in amounts that increase with the decreasing reactivity of alkene 4.

1. Institut für Organische Chemie und Biochemie, TH Darmstadt, Petersenstrasse 22, D-6100 Darmstadt, Germany.
2. Giese, B.; González-Gómes, J. A.; Witzel, T. *Angew. Chem.* **1984**, 96, 51; *Angew. Chem. Intern. Ed. Engl.* **1984**, 23, 69; Giese, B. *Angew. Chem.* **1985**, 97, 555; *Angew. Chem., Intern. Ed. Engl.* **1985**, 24, 553.
3. Kuivila, H. G. *Synthesis* **1970**, 499.

4. Giese, B.; Dupuis, J. *Angew. Chem.* **1983**, *95*, 633; *Angew. Chem., Intern. Ed. Engl.* **1983**, *22*, 622.
5. Dupuis, J.; Giese, B.; Rüegge, D.; Fischer, H.; Korth, H.-G.; Sustmann, R. *Angew. Chem.* **1984**, *96*, 887; *Angew. Chem., Intern. Ed. Engl.* **1984**, *23*, 896.
6. Giese, B.; Gröninger, K. *Tetrahedron Lett.* **1984**, *25*, 2743.
7. Giese, B. *Angew. Chem.* **1983**, *95*, 771; *Angew. Chem., Intern. Ed. Engl.* **1983**, *22*, 753.

Appendix
Chemical Abstracts Nomenclature (Collective Index Number); (Registry Number)

1-Deoxy-2,3,4,6-tetra-O-acetyl-1-(2-cyanoethyl)-α-D-glucopyranose: D-glycero-D-ido-Nonononitrile, 4,8-anhydro-2,3-dideoxy-, 5,6,7,9-tetraacetate (11); (86563-27-1)

Acrylonitrile (8); 2-Propenenitrile (9); (107-13-1)

2,3,4,6-Tetra-O-acetyl-α-D-glucopyranosyl bromide: Glucopyranosyl bromide tetraacetate, α-D- (8); α-D-glucopyranosyl bromide, 2,3,4,6-tetraacetate (9); (572-09-8)

Tributyltin hydride: Stannane, tributyl- (8,9); (688-73-3)

6-BROMO-6-DEOXY HEXOSE DERIVATIVES BY RING-OPENING OF BENZYLIDENE ACETALS WITH N-BROMOSUCCINIMIDE:

METHYL 4-O-BENZOYL-6-BROMO-6-DEOXY-α-D-GLUCOPYRANOSIDE

(Glucopyranoside, methyl 6-bromo-6-deoxy-, 4-benzoate, α-D-)

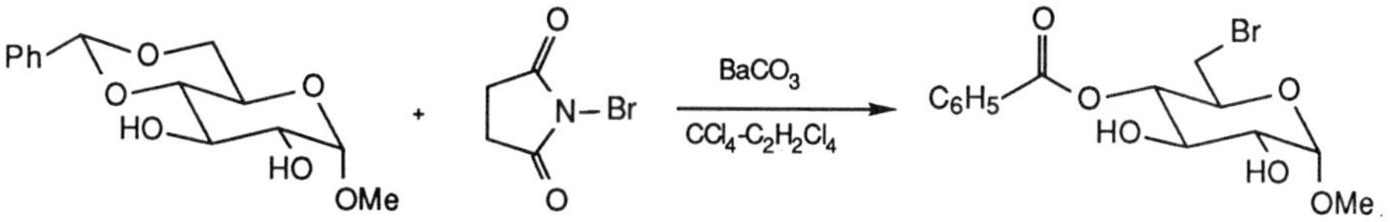

Submitted by S. Hanessian.[1]
Checked by Janice Cammack and James D. White.

1. Procedure

To a suspension containing 20.5 g (72.61 mmol) of methyl 4,6-O-benzylidene-α-D-glucopyranoside (Note 1) in 1 L of carbon tetrachloride and 60 mL of 1,1,2,2-tetrachloroethane (Note 2) are added 15 g (84.27 mmol) of N-bromosuccinimide and 8 g (31.13 mmol) of barium carbonate. The resulting suspension is heated at the reflux temperature of the mixture with mechanical stirring for a period of 2.5 hr and filtered while hot. During the initial period of heating, a reddish orange color develops but fades before termination of the reaction. The yellowish gummy residue in the flask is washed with two 100-mL portions of hot carbon tetrachloride and the filtrate and washings are evaporated under reduced pressure to give a pale yellow oil that is dissolved in 500 mL of ether. The solution is washed with three 60-mL portions of water, then dried over anhydrous sodium sulfate. Evaporation affords a pale yellow oil which crystallizes (Note 3) upon trituration with cold ether to yield 12.1 g of white crystals, mp 120-123°C. A second crop of 1.72 g is obtained from the mother liquors (Note 4). Recrystallization of 1 g of product by dissolution in a minimum volume of acetone and addition of ether, then pentane gives 0.9 g of white crystals, mp 131-132°C; $[\alpha]_D$ +118° ($CHCl_3$, c 1).

2. Notes

1. The preparation of methyl 4,6-O-benzylidene-α-D-glucopyranoside follows essentially the procedure reported.[2]

A mixture of 60 g (0.31 mol) of methyl α-D-glucopyranoside, 45 g of freshly fused and powdered zinc chloride, and 150 mL of benzaldehyde ("practical" grade) is stirred at room temperature for a period of 48 hr. The resulting pale yellow, cloudy reaction mixture is poured slowly, with stirring, into 1.25 L of cold water, stirred for an additional 10 min, and refrigerated overnight. Hexane (75 mL) is added and the resulting mixture is stirred for 0.5 hr to aid in removing excess benzaldehyde. The product is separated on a Buchner funnel, washed twice with 100 mL of cold water, and dried under vacuum at room temperature overnight. Recrystallization from chloroform-ether affords 55 g (63% yield) of analytically pure material, mp 164-165°C.

2. Carbon tetrachloride (spectral grade) was passed through a thick layer of Woelm alumina (neutral). 1,1,2,2-Tetrachloroethane was used as a cosolvent to aid in the dissolution of the starting sugar derivative. In other instances carbon tetrachloride was the solvent.

3. Crystallization did not occur if traces of tetrachloroethane were present. The checkers found it necessary to evaporate at ~ 0.05 mm in a warm water bath for ~ 2 hr to remove residual solvent.

4. Evaporation of the mother liquors and flash column chromatography (350 mL of silica gel; column height 28 cm; eluent 70% ethyl acetate-hexanes, fraction size, 30 mL), give additional (1.87 g) product which was eluted with fractions 25-36; total yield 15.69 g, 60%. (Silica gel, Kieselgel 60; E. Merck AG, Darmstadt, Germany.)

3. Discussion

Halogeno sugar derivatives are versatile intermediates for the preparation of aminodeoxy, deoxy, thio and related analogs.[3] These tranformations are easily achieved in the case of primary halides, which in turn can be prepared by a variety of methods. A number of 6-deoxy and 6-amino-6-deoxy hexoses are components of antibiotics and related natural products.[4,5] Benzylidene acetals of the 1,3-dioxane or 1,3-dioxolane type undergo a ring opening reaction in the presence of N-bromosuccinimide to give the corresponding O-benzoylated bromohydrins.[6] This transformation has been known for a number of years in the carbohydrate series (Hanessian-Hullar reaction),[7] and has been used extensively in synthetic work. In the case of 4,6-O-benzylidene acetals, the products are the 6-bromo-6-deoxy-4-benzoates. Internal acetals of the 1,3-dioxolane type often undergo ring opening to give the two possible regio-isomeric bromo benzoates. The reaction is compatible with a variety of functional and protecting groups (ester, ether, amide, halide, epoxide, etc.). It is also applicable to substrates containing free hydroxyl groups such as the example given above. A unique feature, which arises as a consequence of the nature of the ring opening, is seen in the case of methyl 4,6-O-benzylidene-α-D-galactopyranoside and its derivatives. In this series the benzoate group is found at the C-4 position which has an axial orientation. Hence one achieves halogenation at the primary position as well as an indirect benzoylation of an axial hydroxyl group in the parent sugar. Other applications have been found in amino sugars and nucleosides. Table I illustrates a selection of such ring-opening reactions. The reaction has also been applied with disaccharide acetals.[6,8]

Acknowledgments

The author thanks Mrs. Ani Glamyan for valuable technical assistance.

1. Department of Chemistry, University of Montreal, C.P. 6210, Succursale A., Montreal (Que.), Canada H3C 3V1.
2. Richtmyer, N. K. *Methods Carbohydr. Chem.* **1962**, *1*, 107.
3. See for example, "International Review of Science, Organic Chemistry Series One and Two, Vol. 7, Carbohydrates", Aspinall, G. O., Ed.; Butterworths: London, 1973; 1976.
4. See for example, Umezawa, S. In "International Review of Science, Organic Chemisry Series Two, Vol. 7, Carbohydrates", Aspinall, G. O., Ed.; Butterworths: London, 1976; p. 149.
5. Hanessian, S.; Haskell, T. H. In "The Carbohydrates: Chemistry and Biochemistry", 2nd ed.; Pigman, W.; Horton, D., Eds.; Academic Press: New York, 1970; Vol. IIA, p. 139.
6. See for example, Hanessian, S. *Carbohydr. Res.* **1966**, *2*, 86; Hanessian, S.; Plessas, N. R. *J. Org. Chem.* **1969**, *34*, 1035, 1045, 1053; see also Hanessian, S. *Methods Carbohyd. Chem.* **1972**, *6*, 183.

7. See for example, Guthrie, R. D.; Ferrier, R. J.; How, M. J.; Somers, P. J. In "Carbohydrate Chemistry", Specialist Periodical Report, The Chemical Society, 1968; Vol. 2, p. 3; See also Failla, D. L.; Hullar, T. L.; Siskin, S. B. *J. Chem. Soc., Chem. Commun.* **1966**, 716; Hullar, T. L.; Siskin, S. B. *J. Org. Chem.* **1970**, *35*, 225.
8. Thiem, J.; Horst, K. *Tetrahedron Lett.* **1978**, 4999.
9. Chiba, T.; Haga, M.; Tejima, S. *Chem. Pharm. Bull.* **1975**, *23*, 1283.
10. Baer, H. H.; Georges, F. F. Z. *Can. J. Chem.* **1977**, *55*, 1348.
11. Cheung, T.-M.; Horton, D.; Sorenson, R. J.; Weckerle, W. *Carbohydr. Res.* **1978**, *63*, 77.
12. Florent, J.-C.; Monneret, C.; Khuong-Huu, Q. *Carbohydr. Res.* **1977**, *56*, 301.
13. Ponpipom, M. M.; Hanessian, S. *Can. J. Chem.* **1972**, *50*, 246, 253.

Table I. Reaction of O-Benzylidene Acetals with N-Bromosuccinimide

Starting Acetal	Product(s)	Reference
		5
		5
		5
R = H, Me		5
		5
		5
		9

248

Table I (continued). Reaction of O-Benzylidene Acetals with N-Bromosuccinimide

Starting Acetal	Product(s)	Reference
		10
		11
		5
		12

Appendix
Chemical Abstracts Nomenclature (Collective Index Number);
(Registry Number)

Methyl 4-O-benzoyl-6-bromo-6-deoxy-α-D-glucopyranoside: Glucopyranoside, methyl 6-bromo-6-deoxy-4-benzoate, α-D- (8); glucopyranoside, methyl 6-bromo-6-deoxy, 4-benzoate, α-D- (9); (10368-81-7)

Methyl 4,6-O-benzylidene-α-D-glucopyranoside: Glucopyranoside, methyl 4,6-O-benzylidene-α-D- (8); α-D-glucopyranoside, methyl 4,6-O-(phenylmethylene)- (9); (3162-96-7)

1,1,2,2-Tetrachloroethane: Ethane, 1,1,2,2-tetrachloro- (8,9); (79-34-5)

N-Bromosuccinimide: Succinimide, N-bromo- (8); 2,5-Pyrrolidinedione, 1-bromo- (9); (128-08-5)

Methyl α-D-glucopyranoside: Glucopyranoside, methyl, α-D- (8); α-D-glucopyranoside, methyl (9); (97-30-3)

Unchecked Procedures

Accepted for checking during the period September 1, 1985 through January 1, 1987. An asterisk (*) indicates that the procedure has been subsequently checked.

In accordance with a policy adopted by the Board of Editors, beginning with Volume 50 and further modified subsequently, procedures received by the Secretary and subsequently accepted for checking will be made available upon request to the Secretary, if the request is accompanied by a stamped, self-addressed envelope. (Most manuscripts require 54¢ postage).
Address requests to:

> Professor Jeremiah P. Freeman
> Organic Syntheses, Inc.
> Department of Chemistry
> University of Notre Dame
> Notre Dame, Indiana 46556

It should be emphasized that the procedures which are being made available are unedited and have been reproduced just as they were first received from the submitters. There is no assurance that the procedures listed here will ultimately check in the form available, and some of them may be rejected for publication in *Organic Syntheses* during or after the checking process. For this reason, *Organic Syntheses* can provide no assurance whatsoever that the procedures will work as described and offers no comment as to what safety hazards may be involved. Consequently, more than usual caution should be employed in following the directions in the procedures.

Organic Syntheses welcomes, on a strictly voluntary basis, comments from persons who attempt to carry out the procedures. For this purpose, a Checker's Report form will be mailed out with each unchecked procedure ordered. Procedures which have been checked by or under the supervision of a member of the Board of Editors will continue to be published in the volumes of *Organic Syntheses*, as in the past. It is anticipated that many of the procedures in the list will be published (often in revised form) in *Organic Syntheses* in future volumes.

2350R   From Furfural to 3-Substituted 2-Furanones
J.-A. H. Näsman, Institutionen för Organisk kemi
Åbo Akademi, Akademigatan 1, SF-20500 Åbo 50, Finland

2358R*   Cyclopentanones from Carboxylic Acids via Intramolecular Acylation of Alkylsilanes: 2-Methyl-2-vinylcyclopentanone
I. Kuwajima and H. Urabe, Department of Chemistry, Tokyo Institute of Technology, Meguro, Tokyo 152, Japan

2363R*   Ethyl (E,Z)-2,4-Decadienoate
S. Tsuboi, T. Masuda, and A. Takeda, Department of Synthetic Chemistry, Okayama University, School of Engineering, Okayama, Japan

2386*   Geranyl Pyrophosphate
A. B. Woodside, Z. Huang, and C. D. Poulter, Department of Chemistry, University of Utah, Salt Lake City, UT 84112

2388*   (S)-N,N-Dimethyl-N'(1-$\underline{t}$-butoxy-3-methyl-2-butyl) Formamidine
D. A. Dickman, M. Boes and A. I. Meyers, Department of Chemistry, Colorado State University, Fort Collins, CO 80523

2389   (-)-Salsolidine
A. I. Meyers and M. Boes, Department of Chemistry, Colorado State University, Fort Collins, CO 80523

2391   Diastereoselective Formation of α-Methoxycarbonyl Lactones Through an Intramolecular Diels-Alder-Reaction
L. F. Tietze, G. v. Kiedrowski, and K.-G. Fahlbusch, Institut für Organische Chemie, der Georg August Universität, Tammannstrasse 2, D-3400 Göttingen, Federal Republic of Germany

2393*   6-Diethylphosphonomethyl-2,2-dimethyl-1,3-dioxen-4-one
R. K. Boeckman, Jr., R. B. Perni, J. E. Macdonald, and A. J. Thomas, Department of Chemistry, The University of Rochester, River Station, Rochester, NY 14627

2394R*   Dienophile Activation via Selenosulfonation. 1-(Benzenesulfonyl)-cyclopentene
H.-S. Lin, M. J. Coghlan, and L. A. Paquette, Department of Chemistry, The Ohio State University, Columbus, OH 43210

2395   Prenyltrimethylsilane
C. Biran, J. Gerval, and J. Dunogues, Laboratoire de Chimie Organique du Silicium et de l'Etain, CNRS U.A. 35, Universite de Bordeaux I, 351 Cours de la Liberation, F-33405 Talence, France

2396*   Reduction of Carboxylic Acids to Aldehydes: 6-Oxodecanal
T. Fujisawa and T. Sato, Chemistry Department of Resources, Mie University, Tsu Mie 514, Japan

2398R*   Ethyl α-(Bromomethyl)Acrylate
J. Villieras and M. Rambaud, Synthese Organique Selective Unité associée au C.N.R.S. n° 475, Faculté de Sciences, F-44072 NANTES Cedex, France

2400*   Ketones from Carboxylic Acids and Grignard Reagents: Methyl 6-Oxodecanoate
T. Fujisawa and T. Sato, Chemistry Department of Resources, Mie University, Tsu Mie 514, Japan

2401   Palladium-Catalyzed Reduction of Enol Trifluoromethanesulfonates to Alkenes: Cholesta-3,5-diene
S. Cacchi, E. Morera, and G. Ortar, Institute of Organic Chemistry, University of Rome, via del Castro Laurenziano 9, 00161 Rome, Italy

2402   Cyclopropanation Using an Iron-Containing Methylene Transfer Reagent: 1,1-Diphenylcyclopropane
E. J. O'Connor and P. Helquist, Department of Chemistry, University of Notre Dame, Notre Dame, IN 46556

2403*   (1-Oxo-2-propenyl)trimethylsilane
R. L. Danheiser, D. M. Fink, K. Okano, Y.-M. Tsai, and S. W. Szczepanski, Department of Chemistry, Massachusetts Institute of Technology, Cambridge, MA 02139

2405*   Palladium-Catalyzed Coupling of Acid Chlorides with Organotin Reagents: (E)-Ethyl-4-(4-nitrophenyl)-4-oxo-2-butenoate
A. F. Renaldo, J. W. Labadie, and J. K. Stille, Department of Chemistry, Colorado State University, Ft. Collins, CO 80523

2408R*   3-Hydroxy-1-cyclohexenecarboxaldehyde
T. Hudlicky, M. Neveu, D. Pauley, B. C. Ranu, and H. L. Rigby, Department of Chemistry, Virginia Tech, Blacksburg, VA 24061-0699

2409*   N-Acetyl-N-Phenylhydroxylamine via Catalytic Transfer Hydrogenation of Nitrobenzene using Hydrazine and Rhodium on Carbon
P. W. Oxley, B. M. Adger, M. J. Sasse, and M. A. Forth, Smith Kline and French Research Ltd., Old Powder Mills, Leigh, Tonbridge, Kent, TN11 9AN, England

2410*   N-Benzyl-N-Methoxymethyl-N-(trimethylsilyl)methylamine as an Azomethine Ylide Equivalent
A. Padwa and W. Dent, Department of Chemistry, Emory University, 1515 Pierce Drive, Atlanta, GA 30322

2413*   4-Chlorination of Electron Rich Benzenoid Compounds
J. R. L. Smith, L. C. McKeer, and J. M. Taylor, Department of Chemistry, University of York, York YO1 5DD, England

2414*   Alkylations Using (Propargylium)Dicobalt Hexacarbonyl Salts: Preparation of 2-(1-Methyl-2-propynyl)cyclohexanone
V. Varghese, M. Saha, and K. M. Nicholas, Department of Chemistry, The University of Oklahoma, Norman, OK 73019

2415*   1,4-Functionalization of 1,3-Dienes: 1-Acetoxy-4-diethylamino-2-butene and 1-acetoxy-4-benzylamino-2-butene
J. E. Nyström, T. Rein, and J.-E. Bäckvall, Department of Organic Chemistry, Royal Institute of Technology, S-10044 Stockholm, Sweden

2416   Titanium Mediated Cyclization of a Diyne: 1,2-E,E-bis(Ethylidene)cyclohexane
W. A. Nugent, Central Research & Development Department, E. I. du Pont de Nemours & Co., Experimental Station, Wilmington, DE 19898

2419R[*]   N-tert-Butoxycarbonyl-L-leucinal
O. P. Goel, U. Krolls, M. Stier, and S. Kesten, Warner-Lambert Company, Pharmaceutical Research Division, 2800 Plymouth Road, Ann Arbor, MI 48105

2420   Stereoselective 1,4-Functionalizations of Conjugated Dienes: cis- and trans-1-Acetoxy-4-(dicarbomethoxymethyl)-2-cyclohexene
J.-E. Bäckvall and J. Vagberg, Department of Organic Chemistry, Royal Institute of Technology, S-10044 Stockholm, Sweden

2421   Palladium-Catalyzed Conjugate Reduction of α,β-Unsaturated Carbonyl Compounds with Diphenylsilane and Zinc Chloride Cocatalyst: α,β-Dihydro-β-Ionone
E. Keinan and N. Greenspoon, Department of Organic Chemistry, Weizmann Institute of Science, Rehovot 76100, Israel

2422[*]   Reductive Annulation of Vinyl Sulfones. Bicyclo[4.3.0]non-1-en-4-one
H.-S. Lin and L. A. Paquette, Department of Chemistry, The Ohio State University, Columbus, OH 43210

2424   4-(Trimethylsilyl)-3-butyn-1-ol Tetrahydropyranyl Ether, and (Z)-4-(Trimethylsilyl)-3-buten-1-ol
L. E. Overman, T. C. Malone, and S. F. McCann, Department of Chemistry, University of California, Irvine, CA 92717

2425   Regioselective Synthesis of Tetrahydropyridines. Synthesis of 1-(4-Methoxyphenyl)-1,2,5,6-tetrahydropyridine
L. E. Overman, C. J. Flann, and T. C. Malone, Department of Chemistry, University of California, Irvine, CA 92717

2427[*]   Ethyl 5-Oxo-6-methyl-6-heptenoate from Methacryloyl Chloride and Ethyl 4-Iodobutyrate
Y. Tamaru, H. Ochiai, T. Nakamura, and Z.-i. Yoshida, Department of Synthetic Chemistry, Kyoto University, Yoshida, Sakyo, Kyoto 606, Japan

2428   α-Diphenylmethylsilylation of Ester Enolates: 2-Methyl-2-undecene from Ethyl Decanoate
G. L. Larson, I. Montes de Lopez-Cepero and L. R. Mieles, Department of Chemistry, University of Puerto Rico, Rio Piedras, Puerto Rico 00931

2429[*]   Ethyl α-(Hexahydroazepinylidene-2) Acetate from O-Methyl-caprolactim and Meldrum's Acid
J. P. Celerier, E. Deloisy-Marchalant, G. Lhommet and P. Maitte, Laboratoire de Chimie des Hétérocycles, Université Pierre et Marie Curie, 4, Place Jussieu, 75230 Paris Cedex 05, France

2430* 4,4-Dimethylcyclopent-2-enone
       D. Pauley, F. Anderson and T. Hudlicky, Department of Chemistry,
       Virginia Tech, Blacksburg, VA 24061-0699

2431* 3'-Nitro-1-phenylethanol by Addition of Methyl-triisopropoxy-titanium
       to m-Nitrobenzaldehyde
       R. Imwinkelried and D. Seebach, Laboratorium für Organische Chemie,
       Eidgenössische Technische Hochschule, Universitätstr. 16,
       CH - 8092 Zürich, Switzerland

2433* Condensation of (-)-Dimenthyl Succinate Dianion with 1,ω-Dihalides:
       (-)-Dimenthyl Cyclopropane-(1S,2S)-1,2-dicarboxylate
       K. Furuta, K. Iwanaga, and H. Yamamoto, Department of Applied
       Chemistry, Nagoya University, Furocho, Chikusa, Nagoya 464, Japan

2434* Synthesis of Cyclobutanones via 1-Bromo-1-ethoxycyclopropane: (E)-2-
       (1-Propenyl)cyclobutanone
       S. A. Miller and R. C. Gadwood, Department of Chemistry, Northwestern
       University, Evanston, IL 60201

2438   α-Hydroxylation of Carboxylate Derivatives by Oxygenation of Lithium
       Enolates. Ethyl 1-Hydroxycyclohexanecarboxylate
       H. H. Wasserman, B. H. Lipshutz, and M. C. Corey, Department of
       Chemistry, University of California, Santa Barbara, CA 93106

2440   Selective Cyclopropanation of Perillyl Alcohol: 1-Hydroxymethyl-4-
       (1-methylcyclopropyl)-1-cyclohexene
       K. Maruoka, S. Sakane, and H. Yamamoto, Department of Applied
       Chemistry, Nagoya University, Furocho, Chikusa, Nagoya 464, Japan

2441   Direct Nucleophilic Acylation by the Low Temperature, in situ
       Generation of Acyllithium Reagents; α-Hydroxyketones from Ketones:
       Synthesis of 3-Hydroxy-2,2,3-trimethyloctan-4-one from Pinacolone
       R. Hui and D. Seyferth, Department of Chemistry, Massachusetts
       Institute of Technology, Cambridge, MA 02139

2444   Ethynyl p-Tolyl Sulfone
       L. Waykole and L. A. Paquette, Department of Chemistry, The Ohio
       State University, Columbus, OH 43210

2445   Erythro-directed Reduction of β-Keto Amide: Erythro-1-(3-hydroxy-2-
       methyl-3-phenylpropanoyl)piperidine
       M. Fujita and T. Hiyama, Sagami Chemical Research Center, Nishi-
       Ohnuma 4-4-1, Sagamihara, Kanagawa 229, Japan

2446   Threo-directed Reduction of β-Keto Amide: Threo-1-(hydroxy-2-methyl-
       3-phenylpropanoyl)piperidine
       M. Fujita and T. Hiyama, Sagami Chemical Research Center, Nishi-
       Ohnuma 4-4-1, Sagamihara, Kanagawa 229, Japan

CUMULATIVE AUTHOR INDEX

FOR VOLUME 65

This index comprises the names of contributors to Volume 65 only. For authors to previous volumes, see cumulative indices in Volume 64, which covers Volumes 60 through 64, and Volume 59, which covers Volumes 55 through 59, and either indices in Collective Volumes I through V or single volumes entitled *Organic Syntheses, Collective Volumes*, I, II, III, IV, V, *Cumulative Indices*, edited by R. L. Shriner and R. H. Shriner.

Aslamb, M., **65**, 90

Bercaw, J. E., **65**, 42
Bergman, J., **65**, 146
Bergman, R. G., **65**, 42
Block, E., **65**, 90
Boger, D. L., **65**, 32, 98
Bou, A., **65**, 68
Braxmeier, H., **65**, 159
Brotherton, C. E., **65**, 32
Brown, C. A., **65**, 224
Buter, J., **65**, 150

Corey, E. J., **65**, 166

Davidsen, S. K., **65**, 119
Dupuis, J., **65**, 236

Eliel, E. L., **65**, 215
Enders, D., **65**, 173, 183

Fey, P., **65**, 173, 183
Fowler, K. W., **65**, 108
Frye, S. V., **65**, 215

Georg, G. I., **65**, 32
Giese, B., **65**, 236
Goodwin, G. B. T., **65**, 1
Gross, A. W., **65**, 166

Haese, W., **65**, 26
Hanessian, S., **65**, 243
Hegedus, L. S., **65**, 140

Holmes, A. B., **65**, 52, 61
Hsiao, C.-N., **65**, 135

Imwinkelried, R., **65**, 230

Jadhav, P. K., **65**, 224
Jones, G. E., **65**, 52

Kellogg, R. M., **65**, 150
Kendrick, D. A., **65**, 52
Kipphardt, H., **65**, 173, 183
Kresze, G., **65**, 159
Kuhlmann, H., **65**, 26
Kume, F., **65**, 215
Kuwajima, I., **65**, 17

Le Gal, J. Y., **65**, 47
Lombardo, L., **65**, 81
Lynch, J. E., **65**, 215

Martin, S. F., **65**, 119
McGuire, M. A., **65**, 140
Mickel, S. J., **65**, 135
Miller, M. J., **65**, 135
Mukaiyama, T., **65**, 6
Mullican, M. D., **65**, 98
Munsterer, H., **65**, 159

Nakamura, E., **65**, 17
Narasaka, K., **65**, 6, 12
Nix, M., **65**, 236

Olomucki, M., **65**, 47
Ort, O., **65**, 203

Pericas, M. A., **65**, 68
Phillips, G. W., **65**, 119

Riera, A., **65**, 68
Rodgers, W. B., **65**, 108

Sand, P., **65**, 146
Schiess, M., **65**, 230
Schultze, L. M., **65**, 140
Seebach, D., **65**, 230
Seidler, P. F., **65**, 42
Serratosa, F., **65**, 68
Smith, G. C., **65**, 1
Sporikou, C. N., **65**, 61
Stetter, H., **65**, 26
Stryker, J. M., **65**, 42

Threlkel, R. S., **65**, 42

Walshe, N. D. A., **65**, 1
Wester, R. T., **65**, 108
Woodward, F. E., **65**, 1

Ziegler, F. E., **65**, 108

CUMULATIVE SUBJECT INDEX
FOR VOLUME 65

This index comprises subject matter for Volume 65 only. For subjects in previous volumes, see the cumulative indices in Volume **64**, which covers Volumes **60** through **64**, and Volume **59**, which covers Volumes **55** through **59**, and either the indices in Collective Volumes I through V or the single volume entitled *Organic Syntheses, Collective Volumes I, II, III, IV, V, Cumulative Indices*, edited by R. L. Shriner and R. H. Shriner.

The index lists the names of compounds in two forms. The first is the name used commonly in procedures. The second is the systematic name according to **Chemical Abstracts** nomenclature, accompanied by its registry number in brackets. While the systematic name is indexed separately, it also accompanies the common name. Also included are general terms for classes of compounds, types of reactions, special apparatus, and unfamiliar methods.

Most chemicals used in the procedure will appear in the index as written in the text. There generally will be entries for all starting materials, reagents, intermediates, important by-products, and final products. Entries in capital letters indicate compounds, reactions, or methods appearing in the title of the preparation.

Acetic acid, chloro-, 5-methyl-2-(1-methyl-1-phenylethyl)cyclohexyl ester, [1R-(1α,2β,5α)]-, **65**, 203

Acetic acid ethenyl ester, **65**, 135

Acetic acid, trifluoro-, anhydride, **65**, 12

Acetic acid vinyl ester, **65**, 135

ACETONE TRIMETHYLSILYL ENOL ETHER: SILANE, (ISOPROPENYLOXY)TRIMETHYL-; SILANE, TRIMETHYL[(1-METHYLETHENYL)OXY]-; (1833-53-0), **65**, 1

Acetophenone; Ethanone, 1-phenyl-; (98-86-2), **65**, 6, 119

Acetophenone silyl enol ether: Silane, trimethyl[(1-phenylvinyl)oxy]-; Silane, trimethyl[(1-phenylethenyl)oxy]-; (13735-81-4), **65**, 12

4-ACETOXYAZETIDIN-2-ONE: 2-AZETIDINONE, 4-HYDROXY-ACETATE (ESTER): 2-AZETIDINONE, 4-(ACETYLOXY)-; (28562-53-0), **65**, 135

Acetylene; Ethyne; (74-86-2), **65**, 61

Acrylonitrile; 2-Propenenitrile; (107-13-1), **65**, 236

Aldol reaction, **65**, 6, 12

ALLYLCARBAMATES, **65**, 159

1-AMINO-2-METHOXYMETHYLPYRROLIDINE, (R)-(+)- (RAMP), **65**, 173

1-AMINO-2-METHOXYMETHYLPYRROLIDINE, (S)-(-)- (SAMP), **65**, 173, 183

3-Aminopropylamine: 1,3-Propanediamine; (109-76-2), **65**, 224

Asymmetric synthesis, **65**, 183, 215

AZA-ENE REACTION, **65**, 159

2-AZETIDINONE, 4-(ACETYLOXY)-, **65**, 135

2-AZETIDINONE, 4-HYDROXY-ACETATE (ESTER), **65**, 135

2-AZETIDINONE, 3-METHOXY-1,3-DIMETHYL-4-PHENYL-, **65**, 140

Benzaldehyde; (100-52-7), **65**, 119

Benzaldehyde, 2-bromo-4,5-dimethoxy-, **65**, 108

Benzeneamine, 2-methyl-3-nitro-, **65**, 146

Benzenemethanethiol, **65**, 215

2H-1-BENZOPYRAN-3-CARBOXYLIC ACID, 5,6,7,8-TETRAHYDRO-2-OXO-, METHYL ESTER, **65**, 98

4H-1,3-BENZOXATHIIN, HEXAHYDRO-4,4,7-TRIMETHYL-, **65**, 215

3-Benzyl-5-(2-hydroxyethyl)-4-methyl-1,3-thiazolium chloride; (4568-71-2), **65**, 26

Benzylidenemalononitrile: Malononitrile, benzylidene-; Propanedinitrile, (phenylmethylene)-; (2700-22-3), **65**, 32

Benzyl mercaptan: α-Toluenethiol; Benzenemethanethiol; (100-53-8), **65**, 215

BETA-LACTAM, **65**, 135, 140

BICYCLO[3.1.1]HEPTANE, 6,6-DIMETHYL-2-METHYLENE-, (1S)-, **65**, 224

BICYCLO[3.1.1]HEPT-2-ENE, 2,6,6-TRIMETHYL-, (1S)-, **65**, 224

[1,1'-BIPHENYL]-2,2'-DICARBOXALDEHYDE, 4,4'5,5'-TETRAMETHOXY-, **65**, 108

1,4-BIS(TRIMETHYLSILYL)BUTA-1,3-DIYNE: 2,7-DISILAOCTA-3,5-DIYNE, 2,2,7,7-TETRAMETHYL-; SILANE, 1,3-BUTADIYNE-1,4-DIYLBIS[TRIMETHYL-; (4526-07-2), **65**, 52

1,2-Bis(trimethylsilyloxy)cyclobut-1-ene: Silane, (1-cyclobuten-1,2-ylenedioxy)bis[trimethyl-; Silane, [1-cyclobutene-1,2-diylbis(oxy)]bis[trimethyl-; (17082-61-0), **65**, 17

Boron trifluoride etherate: Ethyl ether, compd. with boron fluoride (BF_3) (1:1); Ethane, 1,1'-oxybis-, compd. with trifluoroborane (1:1); (109-63-7), **65**, 17

2-Bromo-2-butene (cis and trans mixture): 2-Butene, 2-bromo-; (13294-71-8), **65**, 42

1-Bromo-3-chloro-2,2-dimethoxypropane: 2-Propanone, 1-bromo-3-chloro-, dimethyl acetal; Propane, 1-bromo-3-chloro-2,2-dimethoxy-; (22089-54-9), **65**, 2

6-Bromo-3,4-dimethoxybenzaldehyde: Benzaldehyde, 2-bromo-4,5-dimethoxy-; (5392-10-9), **65**, 108

6-Bromo-3,4-dimethoxybenzaldehyde cyclohexylimine: Cyclohexanamine, N-[(2-bromo-4,5-dimethoxyphenyl)methylene]-; (73252-55-8), **65**, 108

BROMOMETHANESULFONYL BROMIDE: METHANESULFONYL BROMIDE, BROMO-; (54730-18-6), **65**, 90

2-(Bromomethyl)-2-(chloromethyl)-1,3-dioxane: 1,3-Dioxane, 2-(bromomethyl)-2-(chloromethyl)-; (60935-30-0), **65**, 32

N-Bromomethylphthalimide: Phthalimide, N-(bromomethyl)-; 1H-Isoindole-1,3-(2H)-dione, 2-(bromomethyl)-; (5332-26-3), **65**, 119

N-BROMOSUCCINIMIDE: SUCCINIMIDE, N-BROMO-; 2,5,-PYRROLIDINEDIONE, 1-BROMO-; (128-08-5), **65**, 243

Butadiyne, **65**, 52

Butane, 1-chloro-, **65**, 61

1-Butanone, 3-hydroxy-3-methyl-1-phenyl-, **65**, 6, 12

2-Butene, 2-bromo-, **65**, 42

2-Butene, 2-methyl-, **65**, 159

2-Buten-1-one, 3-methyl-1-phenyl-, **65**, 12

3-Buten-2-one; (78-94-4), **65**, 26

tert-Butylhydrazine hydrochloride: Hydrazine, tert-butyl, monohydrochloride; Hydrazine, (1,1-dimethylethyl)-, monohydrochloride; (7400-27-3), **65**, 166

N-tert-Butyl-N-tert-octyl-O-tert-butylhydroxylamine: 2-Pentanamine, N-(1,1-dimethylethoxy)-N-(1,1-dimethylethyl)-2,4,4-trimethyl-; (90545-93-0), **65**, 166

TERT-BUTYL-TERT-OCTYLAMINE: 2-PENTANAMINE, N-(1,1-DIMETHYLETHYL)-2,4,4-TRIMETHYL-; (90545-94-1), **65**, 166

Butyllithium: Lithium, butyl-; (109-72-8), **65**, 98, 108, 119

2-BUTYNOIC ACID, 4-CHLORO-, METHYL ESTER, **65**, 47

Carbamic acid, dichloro-, methyl ester, **65**, 159

CARBAMIC ACID, (2-METHYL-2-BUTENYL)-, METHYL ESTER, **65**, 159

Carbamic acid, methyl ester, **65**, 159

Carbonic acid, dicesium salt, **65**, 150

Carbonochloridic acid, methyl ester, **65**, 47

Cesium carbonate: Carbonic acid, dicesium salt; (534-17-8), **65**, 150

CESIUM THIOLATES, **65**, 150

Chiral auxiliary, **65**, 173, 183, 203, 215

1-Chlorobutane: Butane, 1-chloro-; (109-69-3), **65**, 61

2-Chloroethanol; Ethanol, 2-chloro-; (107-07-3), **65**, 150

2-Chloroethyl dichlorophosphate: Phosphorodichloridic acid, 2-chloroethyl ester; (1455-05-6), **65**, 68

Chlorosulfonyl isocyanate: Sulfuryl chloride isocyanate; (1189-71-5), **65**, 135

Chlorotetrolic esters, **65**, 47

Chlorotrimethylsilane: Silane, chlorotrimethyl-; (75-77-4), **65**, 1, 6, 61

Chromium carbonyl, **65**, 140

Chromium hexacarbonyl: Chromium carbonyl (OC-6-11); (13007-92-6), **65**, 140

Chromium, pentacarbonyl(1-methoxyethylidene)-, **65**, 140

Copper(I) bromide (7787-70-4), **65**, 203

Copper(I) chloride - tetramethylethylenediamine complex, **65**, 52

[2 + 2] CYCLOADDITION, **65**, 135

Cyclohexanamine, **65**, 108

Cyclohexanamine, N-[(2-bromo-4,5-dimethoxyphenyl)methylene]-, **65**, 108

Cyclohexanamine, N-[(2-iodo-4,5-dimethoxyphenyl)methylene]-, **65**, 108

Cyclohexane, 1,1-diethoxy-, **65**, 17

CYCLOHEXANE, 1,2-DIMETHYLENE, **65**, 90

CYCLOHEXANE, 1,2-BIS(METHYLENE)-, **65**, 90

CYCLOHEXANE, 5-METHYL-1-METHYLENE-2-(1'-METHYLETHYL)-, R,R-, **65**, 81

CYCLOHEXANEBUTANOIC ACID, γ-OXO-, ETHYL ESTER, **65**, 17

Cyclohexanol, 2-(1-mercapto-1-methylethyl)-5-methyl-, [(1R-(1α,2α,5α)]-, **65**, 215

Cyclohexanol, 5-methyl-2-(1-methylethyl)-, [1S-(1α,2β,5β)], **65**, 81

CYCLOHEXANOL, 5-METHYL-2-(1-METHYL-1-PHENYLETHYL)-, [1R-(1α,2β,5α)]-, **65**, 203

Cyclohexanone; (108-94-1), **65**, 98

Cyclohexanone diethyl ketal; Cyclohexane, 1,1-diethoxy-; (1670-47-9), **65**, 17

Cyclohexanone, 5-methyl-2-(1-methylethyl)-, (2R-cis)-, **65**, 81

CYCLOHEXANONE, 5-METHYL-2-(1-METHYLETHYLIDENE)-, (R)-, **65**, 203, 215

Cyclohexanone, 5-methyl-2-(1-methyl-1-phenylethyl)-, (2R-trans)-, **65**, 203

Cyclohexanone, 5-methyl-2-(1-methyl-1-phenylethyl)-, (2S-cis)-, **65**, 203

Cyclohexanone, 5-methyl-2-[1-methyl-1-(phenylmethylthio)ethyl]-, (2R-trans)-; (79563-58-9); (2S-cis)-; (79618-04-5), **65**, 215

Cyclohexene, 1-methyl-, **65**, 90

Cyclohexylamine; Cyclohexanamine; (108-91-8), **65**, 108

1,3-CYCLOPENTADIENE, 1,2,3,4,5-PENTAMETHYL-, **65**, 42

2-CYCLOPENTEN-1-ONE, 3-METHYL-2-PENTYL-, **65**, 26

CYCLOPROPENONE 1,3-PROPANEDIOL KETAL: 4,8-DIOXASPIRO[2.5]OCT-1-ENE; (60935-21-9), **65**, 32

1-DEOXY-2,3,4,6-TETRA-O-ACETYL-1-(2-CYANOETHYL)-α-D-GLUCOPYRANOSE:
D-GLYCERO-D-IDO-NONONONITRILE, 4,8-ANHYDRO-2,3-DIDEOXY-,
5,6,7,9-TETRAACETATE; (86563-27-1), **65**, 236

DIALKOXYACETYLENES, **65**, 68

Dibromomethane: Methane, dibromo-; (74-95-3), **65**, 81

1,2-Di-tert-butoxy-1-chloroethene, (E)-: Propane, 2,2'-[(1-chloro-1,2-ethenediyl)bis(oxy)]bis[2-methyl-, (E); (70525-93-8), **65**, 58

1,2-Di-tert-butoxy-1,2-dichloroethane, dl-: Propane, 2,2'-[(1,2-dichloro-1,2-ethanediyl)]bis(oxy)bis[2-methyl-, (R^*,R^*)-(±)-; (68470-80-4), **65**, 68

1,2-Di-tert-butoxy-1,2-dichloroethane, meso-: Propane, 2,2'-[(1,2-dichloro-1,2-ethanediyl)bis(oxy)]bis[2-methyl-, (R^*,S^*)-; (68470-81-5), **65**, 68

2,3-Di-tert-butoxy-1,4-dioxane, cis-: 1,4-Dioxane, 2,3-bis(1,1-dimethylethoxy)-, cis-; (68470-78-0), **65**, 68

2,3-Di-tert-butoxy-1,4-dioxane, trans-: 1,4-Dioxane, 2,3-bis(1,1-dimethylethoxy)-, trans-; (68470-79-1), **65**, 68

DI-TERT-BUTOXYETHYNE: PROPANE, 2,2'-[1,2-ETHYNEDIYLBIS(OXY)]BIS[2-METHYL-; (66478-63-5), **65**, 68

2,3-Dichloro-1,4-dioxane, trans-: 1,4-Dioxane, 2,3-dichloro-, trans-; (3883-43-0), **65**, 68

5,5-Dicyano-4-phenylcyclopent-2-enone 1,3-propanediol ketal: 6,10-Dioxaspiro[4.5]dec-3-ene-1,1-dicarbonitrile, 2-phenyl-; (88442-12-0), **65**, 32

Diethyl aminomethylphosphonate: Phosphonic acid, (aminomethyl)-, diethyl ester; (50917-72-1), **65**, 119

DIETHYL N-BENZYLIDENEAMINOMETHYLPHOSPHONATE: PHOSPHONIC ACID, [[(PHENYLMETHYLENE)AMINO]METHYL]-, DIETHYL ESTER; (50917-73-2), **65**, 119

Diethyl isocyanomethylphosphonate: Phosphonic acid, (isocyanomethyl)-, diethyl ester; (41003-94-5), **65,** 119

Diethyl oxalate: Oxalic acid, diethyl ester; Ethanedioic acid, diethyl ester; (95-92-1), **65,** 146

Diethyl phthalimidomethylphosphonate: Phosphonic acid, (phthalimidomethyl)-, diethyl ester; Phosphonic acid, [(1,3-dihydro-1,3-dioxo-2H-isoindol-2-yl)-methyl]-, diethyl ester; (33512-26-4), **65,** 119

DIHYDROJASMONE, **65,** 26

Diisopropylamine; 2-Propanamine, N-(1-methylethyl)-; (108-18-9), **65,** 98

DIISOPROPYL (2S,3S)-2,3-O-ISOPROPYLIDENETARTRATE: 1,3-DIOXOLANE-4,5-DICARBOXYLIC ACID, 2,2-DIMETHYL-, BIS(1-METHYLETHYL) ESTER, (4R-TRANS)-; (81327-47-1), **65,** 230

1,1-Dimethoxyethylene: Ethene, 1,1-dimethoxy-; (922-69-0), **65,** 98

4-(N,N-Dimethylamino)pyridine: Pyridine, 4-(dimethyamino)-; 4-Pyridinamine, N,N-dimethyl-; (1122-58-3), **65,** 12

Dimethyl diazomethylphosphonate: Phosphonic acid, (diazomethyl)-, dimethyl ester; (27491-70-9), **65,** 119

3,3-DIMETHYL-1,5-DIPHENYLPENTANE-1,5-DIONE: 1,5-PENTANEDIONE, 3,3-DIMETHYL-1,5-DIPHENYL-; (42052-44-8), **65,** 12

1,2-DIMETHYLENECYCLOHEXANE: CYCLOHEXANE, 1,2-DIMETHYLENE; CYCLOHEXANE, 1,2-BIS(METHYLENE)-; (2819-48-9), **65,** 90

Dimethyl (2S,3S)-2,3-O-isopropylidenetartrate: 1,3-Dioxolane-4,5-dicarboxylic acid, 2,2-dimethyl-, dimethyl ester, (4R-trans)- or (4S-trans)-; (37031-29-1) or (37031-30-4), **65,** 230

Dimethyl methoxymethylenemalonate: Malonic acid, (methoxymethylene)-, dimethyl ester; Propanedioic acid, (methoxymethylene)-, dimethyl ester; (22398-14-7), **65,** 98

1,3-DIMETHYL-3-METHOXY-4-PHENYLAZETIDINONE: 2-AZETIDINONE, 3-METHOXY-
1,3-DIMETHYL-4-PHENYL-; (82918-98-7), **65**, 140

DMAP, **65**, 12

1,3-Dioxane, 2-(bromomethyl)-2-(chloromethyl)-, **65**, 32

1,4-Dioxane, 2,3-dichloro-, trans-, **65**, 68

1,4-Dioxane, 2,3-bis(1,1-dimethylethoxy)-, trans-, **65**, 68

1,4-Dioxane, 2,3-bis(1,1-dimethylethoxy)-, cis-, **65**, 68

6,10-Dioxaspiro[4.5]dec-3-ene-1,1-dicarbonitrile, 2-phenyl-, **65**, 32

4,8-DIOXASPIRO[2.5]OCT-1-ENE, **65**, 32

p-Dioxino[2,3,-b]-p-dioxin, hexahydro, **65**, 68

[1,4]-Dioxino[2,3-b]-1,4-dioxin, hexahydro, **65**, 68

1,3-Dioxolane-4,5-dicarboxylic acid, 2,2-dimethyl-, dimethyl ester,
(4R-trans)- or (4S-trans)-, **65**, 230

1,3-DIOXOLANE-4,5-DICARBOXYLIC ACID, 2,2-DIMETHYL-, BIS(1-METHYLETHYL) ESTER,
(4R-TRANS)-, **65**, 230

2,7-DISILAOCTA-3,5-DIYNE, 2,2,7,7-TETRAMETHYL-, **65**, 52

3,7-Dithianonane-1,9-diol: Ethanol, 2,2'-(trimethylenedithiol)di-;
Ethanol, 2,2'-[1,3-propanediylbis(thio)]bis-;
(16260-48-3), **65**, 150

3,7-Dithianonane-1,9-dithiol: Ethanethiol, 2,2'-(trimethylenedithio)di-;
Ethanethiol, 2,2'-[1,3-propanediylbis(thio)]bis-; (25676-62-4), **65**, 150

Ethane, 1,1',1"-[methylidynetris(oxy)]tris-, **65**, 146

Ethane, 1,1'-oxybis-, compd. with trifluoroborane (1:1), **65**, 17

Ethanedioic acid, diethyl ester, **65**, 146

Ethanethiol, 2,2'-[1,3-propanediylbis(thio)]bis-, **65**, 150

Ethanethiol, 2,2'-(trimethylenedithio)di-, **65**, 150

Ethanol, 2-chloro-, **65**, 150

Ethanol, 2,2'-[1,3-propanediylbis(thio)]bis-, **65**, 150

Ethanol, 2,2'-(trimethylenedithiol)di-, **65**, 150

Ethanone, 1-phenyl-, **65**, 6, 119

Ethene, 1,1-dimethoxy-, **65**, 98

ETHYL 4-CYCLOHEXYL-4-OXOBUTANOATE: CYCLOHEXANEBUTANOIC ACID, γ-OXO-, ETHYL ESTER; (54966-52-8), **65**, 17

Ethyl ether, compd. with boron fluoride (BF_3) (1:1), **65**, 17

Ethyl N-(2-methyl-3-nitrophenyl)formimidate, **65**, 146

Ethylenediaminetetraacetic acid, tetrasodium salt: Glycine, N,N'-1,2-ethanediylbis[N-(carboxymethyl)]-, tetrasodium salt, trihydrate; (67401-50-7), **65**, 166

Ethyne, **65**, 61

Formic acid, chloro-, methyl ester, **65**, 47

Glucopyranoside, methyl, α-D-, **65**, 243

Glucopyranoside, methyl 4,6-O-benzylidene-, α-D-, **65**, 243

GLUCOPYRANOSIDE, METHYL 6-BROMO-6-DEOXY, 4-BENZOATE, α-D-, **65**, 243

Glucopyranoside, methyl 4,6-O-(phenylmethylene)-, α-D-, **65**, 243

Glucopyranosyl bromide, 2,3,4,6-tetraacetate, α-D-, **65**, 236

Glucopyranosyl bromide tetraacetate, α-D-, **65**, 236

GLYCERO-D-IDO-NONONONITRILE, 4,8-ANHYDRO-2,3-DIDEOXY-, 5,6,7,9-TETRAACETATE, α-D-, **65**, 236

Glycine, N,N'-1,2-ethanediylbis[N-(carboxymethyl)]-, tetrasodium salt, trihydrate, **65**, 166

2,5-Heptadien-4-ol, 3,4,5-trimethyl-, **65**, 42

Heptanal; (111-71-7), **65**, 26

3-HEPTANONE, 4-METHYL-, (S)-, **65**, 183

HEXAHYDRO-4,4,7-TRIMETHYL-4H-1,3-BENZOXATHIIN: 4H-1,3-BENZOXATHIIN, HEXAHYDRO-4,4,7-TRIMETHYL-; (59324-06-0), **65**, 215

Hydrazine, (1,1-dimethylethyl)-, monohydrochloride, **65**, 166

Hydrazine, tert-butyl, monohydrochloride, **65**, 166

Hydrogen peroxide; (7722-84-1), **65**, 166

3-HYDROXY-3-METHYL-1-PHENYL-1-BUTANONE: 1-BUTANONE, 3-HYDROXY-3-METHYL-1-PHENYL-; (43108-74-3), **65**, 6, 12

N-Hydroxymethylphthalimide: Phthalimide, N-(hydroxymethyl)-; 1H-Isoindole-1,3-(2H)-dione, 2-(hydroxymethyl)- (118-29-6), **65**, 119

INDOLE, 4-NITRO-, **65**, 146

INVERSE ELECTRON DEMAND DIELS-ALDER, **65**, 98

6-Iodo-3,4-dimethoxybenzaldehyde cyclohexylimine: Cyclohexanamine, N-[(2-iodo-4,5-dimethoxyphenyl)methylene]-; (61599-78-8), **65**, 108

Isoindole-1,3-(2H)-dione, 2-(bromomethyl)-, 1H-, **65**, 119

Isoindole-1,3-(2H)-dione, 2-(hydroxymethyl)-, 1H-, **65**, 119

Isomenthol, (+)-: Cyclohexanol, 5-methyl-2-(1-methylethyl)-, [1S-(1α,2β,5β)]-; (23283-97-8), **65**, 81

Isomenthone, (+)-: Cyclohexanone, 5-methyl-2-(1-methylethyl)-, (2R-cis)-; (1196-31-2), **65**, 81

Isopropyl alcohol, titanium (4+) salt, **65**, 230

Isopropylideneacetophenone: 2-Buten-1-one, 3-methyl-1-phenyl-; (5650-07-7), **65**, 12 β-Lactams, **65**, 140

Lead dioxide: Lead oxide; (1309-60-0), **65**, 166

Lead oxide, **65**, 166

Lithiobutadiyne, **65**, 52

Lithium, butyl-, **65**, 98, 108, 119

Lithium diisopropylamide, **65**, 98

Lithium, methyl-, **65**, 47, 140

MACROCYCLIC SULFIDES, **65**, 150

Malonic acid, (methoxymethylene)-, dimethyl ester, **65**, 98

Malononitrile, benzylidene-, **65**, 32

p-MENTH-4-(8)-EN-3-ONE, (R)-(+)-, **65**, 203, 215

2-(1-Mercapto-1-methylethyl)-5-methylcyclohexanol: Cyclohexanol-2-
 (1-mercapto-1-methylethy)-5-methyl-, [1R-(1α,2α,5α)]-;
 (79563-68-1); [1R-(1α,2β,5α)]-; (79563-59-0); [1S-(1α,2α,5β)]-;
 (79563-67-0), **65**, 215

Methanamine, N-(phenylmethylene)-, **65**, 140

Methane, dibromo-, **65**, 81

METHANESULFONYL BROMIDE, BROMO-, **65**, 90

N^1N^2-Bis(methoxycarbonyl)sulfur diimide: Sulfur diimide, dicarboxy-,
 dimethyl ester; (16762-82-6), **65**, 159

6-Methoxy-7-methoxycarbonyl-1,2,3,4-tetrahydronaphthalene: 2-Naphthalene-
 carboxylic acid, 5,6,7,8-tetrahydro-3-methoxy-, methyl ester;
 (78112-34-2), **65**, 98

Methylamine, N-benzylidene-, **65**, 140

METHYL 4-O-BENZOYL-6-BROMO-6-DEOXY-α-D-GLUCOPYRANOSIDE:
   GLUCOPYRANOSIDE, METHYL 6-BROMO-6-DEOXY, 4-BENZOATE,
   α-D-; (10368-81-7), **65**, 243

Methyl 4,6-O-benzylidene-α-D-glucopyranoside: Glucopyranoside, methyl 4,6-O-benzylidene-α-D-; α-D-glucopyranoside, methyl 4,6-O-(phenylmethylene)-;
   (3162-96-7), **65**, 243

N-Methylbenzylideneimine: Methylamine, N-benzylidene-; Methanamine,
   N-(phenylmethylene)-; (622-29-7), **65**, 140

2-Methyl-2-butene: 2-Butene, 2-methyl-; (513-35-9), **65**, 159

Methyl carbamate: Carbamic acid, methyl ester; (598-55-0), **65**, 159

METHYL 4-CHLORO-2-BUTYNOATE: 2-BUTYNOIC ACID, 4-CHLORO-, METHYL ESTER;
   (41658-12-2), **65**, 47

Methyl chloroformate: Formic acid, chloro-, methyl ester;
   Carbonochloridic acid, methyl ester; (79-22-1), **65**, 47

1-Methylcyclohexene: Cyclohexene, 1-methyl-; (591-49-1), **65**, 90

Methyl N,N-dichlorocarbamate: Carbamic acid, dichloro-, methyl ester;
   (16487-46-0), **65**, 159

Methyl α-D-glucopyranoside: Glucopyranoside, methyl, α-D-;
   α-D-glucopyranoside, methyl; (97-30-3), **65**, 243

4-METHYL-3-HEPTANONE, (S)-(+)-: 3-HEPTANONE, 4-METHYL-, (S)-;
   (51532-30-0), **65**, 183

4-Methyl-3-heptanone SAMP-hydrazone, (S)-(+)-: 1-Pyrrolidinamine,
   N-(1-ethyl-2-methylpentylidene)-2-(methoxymethyl)-,
   [S-[R*,R*-(Z)]]-; (69943-24-4), **65**, 183

Methyllithium: Lithium, methyl-; (917-54-4), **65**, 47, 140

[(Methyl)(methoxy)carbene]pentacarbonyl chromium(0): Chromium,
   pentacarbonyl(1-methoxyethylidene)-, (OC-6-21)-; (20540-69-6), **65**, 140

METHYL N-(2-METHYL-2-BUTENYL)CARBAMATE: CARBAMIC ACID, (2-METHYL-2-BUTENYL)-, METHYL ESTER; (86766-65-6), **65**, 159

5-Methyl-2-(1-methyl-1-phenylethyl)cyclohexanone, (2R,5R)-: Cyclohexanone, 5-methyl-2-(1-methyl-1-phenylethyl)-, (2R-trans)-; (57707-92-3), **65**, 203

5-Methyl-2-(1-methyl-1-phenylethyl)cyclohexanone, (2S,5R)-: Cyclohexanone, 5-methyl-2-(1-methyl-1-phenylethyl)-, (2S-cis)-; (65337-06-6), **65**, 203

5-Methyl-2-(1-methyl-1-phenylethyl)cyclohexyl chloroacetate, (1R,2S,5R)-: Acetic acid, chloro-, 5-methyl-2-(1-methyl-1-phenylethyl)cyclohexyl ester, [1R-(1α,2β,5α)]-; (71804-27-8), **65**, 203

5-Methyl-2-[1-methyl-1-(phenylmethylthio)ethyl]cyclohexanone, cis- and trans-; **65**, 215

5-Methyl-2-(1-methyl-1-thioethyl)cyclohexanol, **65**, 215

2-Methyl-3-nitroaniline: o-Toluidine, 3-nitro-; Benzeneamine, 2-methyl-3-nitro-; (603-83-8), **65**, 146

METHYL 2-OXO-5,6,7,8-TETRAHYDRO-2H-1-BENZOPYRAN-3-CARBOXYLATE: 2H-1-BENZOPYRAN-3-CARBOXYLIC ACID, 5,6,7,8-TETRAHYDRO-2-OXO-, METHYL ESTER; (85531-80-2), **65**, 98

3-METHYL-2-PENTYL-2-CYCLOPENTEN-1-ONE: 2-CYCLOPENTEN-1-ONE, 3-METHYL-2-PENTYL-; (1128-08-1), **65**, 26

2-METHYL-2-PHENYL-4-PENTENAL: 4-PENTENAL, 2-METHYL-2-PHENYL-; (24401-39-6), **65**, 119

METHYLENATION OF CARBONYL COMPOUNDS, **65**, 81

3-METHYLENE-CIS-P-MENTHANE, (+)-: (CYCLOHEXANE, 5-METHYL-1-METHYLENE-2-(1'-METHYLETHYL)-, R,R-), **65**, 81

2-Naphthalenecarboxylic acid, 5,6,7,8-tetrahydro-3-methoxy-, methyl ester, **65**, 98

4-NITROINDOLE: INDOLE, 4-NITRO-; (4769-97-5), **65**, 146

1-Nitroso-2-methoxymethylpyrrolidine, (S)-: Pyrrolidine, 2-(methoxymethyl)-1-nitroso-, (S)-; (60096-50-6), **65**, 183

Nitroso-tert-octane: Pentane, 2,2,4-trimethyl-4-nitroso-; (31044-98-1), **65**, 166

tert-Octylamine: 2-Pentanamine, 2,4,4-trimethyl-; (107-45-9), **65**, 166

N-tert-Octyl-O-tert-butylhydroxylamine: 2-Pentanamine, N-(1,1-dimethylethoxy)-2,4,4-trimethyl-; (68295-32-9), **65**, 166

Orthoformic acid, triethyl ester, **65**, 146

Oxalic acid, diethyl ester, **65**, 146

1,3-OXATHIANE, **65**, 215

Oxonium, trimethyl-, tetrafluoroborate (1-), **65**, 140

Ozone (10028-15-6), **65**, 183

Paraformaldehyde: Poly(oxymethylene); (9002-81-7), **65**, 215

1,2,3,4,5-PENTAMETHYLCYCLOPENTADIENE: 1,3-CYCLOPENTADIENE, 1,2,3,4,5-PENTAMETHYL-; (4045-44-7), **65**, 42

2-Pentanamine, N-(1,1-dimethylethoxy)-N-(1,1-dimethylethyl)-2,4,4-trimethyl-, **65**, 166

2-Pentanamine, N-(1,1-dimethylethoxy)-2,4,4-trimethyl-, **65**, 166

2-PENTANAMINE, N-(1,1-DIMETHYLETHYL)-2,4,4-TRIMETHYL-, **65**, 166

2-Pentanamine, 2,4,4-trimethyl-, **65**, 166

1,5-PENTANEDIONE, 3,3-DIMETHYL-1,5-DIPHENYL-, **65**, 12

Pentane, 2,2,4-trimethyl-4-nitroso-, **65**, 166

3-Pentanone SAMP-hydrazone: 1-Pyrrolidinamine, N-(1-ethylpropylidene)-2-(methoxymethyl)-, (S)-; (59983-36-7), **65**, 183

4-PENTENAL, 2-METHYL-2-PHENYL-, **65**, 119

8-PHENYLMENTHOL, (-)-: CYCLOHEXANOL, 5-METHYL-2-(1-METHYL-1-PHENYLETHYL)-, [1R-(1α,2β,5a)]-; (65253-04-5), **65**, 203

2-Phenyl-N-(phenylmethylene)-1-propen-1-amine: 1-Propen-1-amine, 2-phenyl-N-(phenylmethylene)-; (64244-34-4), **65**, 119

1-Phenyl-1-trimethylsiloxyethylene: Silane, trimethyl[(1-phenylvinyl)oxy]-; Silane, trimethyl[(1-phenylethenyl)oxy]-; (13735-81-4), **65**, 6

Phosphonic acid, (aminomethyl), diethyl ester, **65**, 119

Phosphonic acid, (diazomethyl)-, dimethyl ester, **65**, 119

Phosphonic acid, [(1,3-dihydro-1,3-dioxo-2H-isoindol-2-yl)methyl]-, diethyl ester, **65**, 119

Phosphonic acid, (isocyanomethyl)-, diethyl ester, **65**, 119

PHOSPHONIC ACID [[(PHENYLMETHYLENE)AMINO]METHYL]-, DIETHYL ESTER, **65**, 119

Phosphonic acid, (phthalimidomethyl)-, diethyl ester, **65**, 119

Phosphorodichloridic acid, 2-chloroethyl ester, **65**, 68

Phosphorous acid, triethyl ester, **65**, 108, 119

Phthalimide, N-(bromoethyl)-, **65**, 119

Phthalimide, N-(hydroxymethyl)-, **65**, 119

PINENE, (-)-α-: 2-PINENE, (1S,5S)-(-); BICYCLO[3.1.1]HEPT-2-ENE, 2,6,6-TRIMETHYL-, (1S)-; (7785-26-4), **65**, 224

PINENE, (-)-β-: BICYCLO[3.1.1]HEPTANE, 6,6-DIMETHYL-2-METHYLENE-, (1S)-; (18172-67-3), **65**, 224

Poly(oxymethylene), **65**, 215

Potassium 3-aminopropylamide (KAPA), **65**, 224

Potassium hydride; (7693-26-7), **65**, 224

Proline, D-; (344-25-2), **65**, 173

Proline, L-; (147-85-3), **65**, 173

2-Propanamine, N-(1-methylethyl)-, **65**, 98

Propane, 1-bromo-3-chloro-2,2-dimethoxy-, **65**, 32

Propane, 2,2'-[(1-chloro-1,2-ethenediyl)bis(oxy)]bis[2-methyl-, (E)-, **65**, 68

Propane, 2,2'-[(1,2-dichloro-1,2-ethanediyl)]bis(oxy)bis[2-methyl-, (R^*,R^*)-(±)-, **65**, 68

Propane, 2,2'-[(1,2-dichloro-1,2-ethanediyl)bis(oxy)]bis[2-methyl-, (R^*,S^*)-, **65**, 68

PROPANE, 2,2'-[1,2-ETHYNEDIYLBIS(OXY)]BIS[2-METHYL-, **65**, 68

1,3-Propanediamine, **65**, 224

Propanedinitrile, (phenylmethylene)-, **65**, 32

Propanedioic acid, (methoxymethylene)-, dimethyl ester, **65**, 98

1,3-Propanediol; (504-63-2), **65**, 32

1,3-Propanedithiol; (109-80-8), **65**, 150

2-Propanol, titanium (4+) salt, **65**, 230

2-Propanone, 1-bromo-3-chloro-, dimethyl acetal, **65**, 32

PROPARGYL CHLORIDE: PROPYNE, 3-CHLORO-; 1-PROPYNE, 3-CHLORO-; (624-65-7), **65**, 47

1-Propen-1-amine, 2-phenyl-N-(phenylmethylene)-, **65**, 119

2-Propenenitrile, **65**, 236

1-PROPYNE, 3-CHLORO-, **65**, 47

PULEGONE, (+): p-MENTH-4-(8)-EN-3-ONE, (R)-(+)-; CYCLOHEXANONE, 5-METHYL-2-(1-METHYLETHYLIDENE)-, (R)-; (89-82-7), **65**, 203, 215

4-Pyridinamine, N,N-dimethyl-, **65**, 12

Pyridine, 4-(dimethylamino)-, **65**, 12

1-Pyrrolidinamine, N-(1-ethyl-2-methylpentylidene)-2-(methoxymethyl)-,
    [S-[R*,R*-(Z)]]-, **65**, 183
1-Pyrrolidinamine, N-(1-ethylpropylidene)-2-(methoxymethyl)-, (S)-, **65**, 183
1-Pyrrolidinamine, 2-(methoxymethyl)-, (R)-(+)-; (72748-99-3), **65**, 173
1-Pyrrolidinamine, 2-(methoxymethyl)-, (S)-(-)-; (59983-30-0), **65**, 173, 183
Pyrrolidine, 2-(methoxymethyl)-, (S)-(+)-; (63126-47-6), **65**, 173
Pyrrolidine, 2-(methoxymethyl)-1-nitroso-, (S)-, **65**, 183
1-Pyrrolidinecarboxaldehyde, 2-(hydroxymethyl)-, (S)-(-)-; (55456-46-7),
    65, 173
1-Pyrrolidinecarboxaldehyde, 2-(methoxymethyl)-, (S)-(-)-; (63126-45-4),
    65, 173
2,5-PYRROLIDINEDIONE, 1-BROMO-, **65**, 243
2-Pyrrolidinemethanol, (S)-(+)-; (23356-96-9), **65**, 173

Ramberg-Bäcklund reaction, **65**, 90
RAMP: (R)-1-Amino-2-methoxymethylpyrrolidine: 1-Pyrrolidinamine,
    2-(methoxymethyl)-, (R)-; **65**, 173, 183
RING EXPANSION, **65**, 17

SAMP: (S)-1-Amino-2-methoxymethylpyrrolidine: 1-Pyrrolidinamine,
    2-(methoxymethyl)-, (S)-; (59983-39-0), **65**, 173, 183
SILANE, 1,3-BUTADIYNE-1,4-DIYLBIS[TRIMETHYL-, **65**, 52
Silane, chlorotrimethyl-, **65**, 61, 1
Silane, (1-cyclobuten-1,2-ylenedioxy)bis[trimethyl-, **65**, 17
Silane, [1-cyclobutene-1,2-diylbis(oxy)]bis[trimethyl-, **65**, 17
Silane, ethynyltrimethyl-, **65**, 52, 61
SILANE, (ISOPROPENYLOXY)TRIMETHYL-, **65**, 1

SILANE, TRIMETHYL[(1-METHYLETHENYL)OXY]-, **65**, 1

Silane, trimethyl[(1-phenylethyl)oxy]-, **65**, 6, 12

Silane, trimethyl[(1-phenylvinyl)oxy]-, **65**, 6, 12

Sodium naphthalenide, **65**, 166

Sodium tungstate dihydrate: Tungstic acid, disodium salt, dihydrate; (10213-10-2), **65**, 166

SPIRO[4.5]DECAN-1,4-DIONE; (39984-92-4), **65**, 17

Stannane, tetrachloro-, **65**, 17

Stannane, tributyl-, **65**, 236

STETTER REACTION, **65**, 26

SUCCINIMIDE, N-BROMO-, **65**, 243

Sulfur chloride, **65**, 159

Sulfur dichloride: Sulfur chloride; (10545-99-0), **65**, 159

Sulfur diimide, dicarboxy-, dimethyl ester, **65**, 159

Sulfuryl chloride isocyanate, **65**, 135

2,3,4,6-Tetra-O-acetyl-α-D-glucopyranosyl bromide: Glucopyranosyl bromide tetraacetate, α-D-; α-D-glucopyranosyl bromide, 2,3,4,6-tetraacetate; (572-09-8), **65**, 236

Tetraisopropyl titanate: Isopropyl alcohol, titanium (4+) salt; 2-Propanol, titanium (4+) salt; (546-68-9), **65**, 230

4,5,4',5'-TETRAMETHOXY-1,1'-BIPHENYL-2,2'-DICARBOXALDEHYDE: [1,1'-BIPHENYL]-2,2'-DICARBOXALDEHYDE, 4,4',5,5'-TETRAMETHOXY-; (29237-14-7), **65**, 108

2,5,7,10-Tetraoxabicyclo[4.4.0]decane, cis-: p-Dioxino[2,3,-b]-p-dioxin, hexahydro-; [1,4]-Dioxino[2,3-b]-1,4-dioxin, hexahydro; (4362-05-4), **65**, 68

1,4,8,11-TETRATHIACYCLOTETRADECANE; (24194-61-4), **65**, 150

Thiourea: Urea, thio-; (62-56-6), **65**, 150

Tin chloride, **65**, 17

Tin tetrachloride: Tin chloride; Stannane, tetrachloro-; (7646-78-8), **65**, 17

Titanium chloride, **65**, 81, 6

Titanium tetrachloride: Titanium chloride; (7550-45-0), **65**, 81, 6

α-Toluenethiol, **65**, 215

o-Toluidine, 3-nitro-, **65**, 146

Transesterification, **65**, 230

Tributyltin hydride: Stannane, tributyl-; (688-73-3), **65**, 236

Triethyl orthoformate: Orthoformic acid, triethyl ester; Ethane, 1,1',1"-[methylidynetris(oxy)]tris-; (122-51-0), **65**, 146

Triethyl phosphite: Phosphorous acid, triethyl ester; (122-52-1), **65**, 119, 108

Trifluoroacetic anhydride: Acetic acid, trifluoro-, anhydride; (407-25-0), **65**, 12

3,4,5-Trimethyl-2,5-heptadien-4-ol: 2,5-Heptadien-4-ol, 3,4,5-trimethyl-; (64417-15-8), **65**, 42

Trimethyloxonium tetrafluoroborate: Oxonium, trimethyl-, tetrafluoroborate (1-); (420-37-1), **65**, 140

TRIMETHYLSILYLACETYLENE: SILANE, ETHYNYLTRIMETHYL-; (1066-54-2), **65**, 52, 61

Trithiane, Sym-: S-Trithiane; 1,3,5-Trithiane; (291-21-4), **65**, 90

Tungstic acid, disodium salt, dihydrate, **65**, 166

ULLMANN REACTION, **65**, 108

2,5-Undecanedione; (7018-92-0), **65**, 26

Urea, thio-, **65**, 150

Vibro-mixer, **65**, 52

Vinyl acetate: Acetic acid vinyl ester; Acetic acid ethenyl ester; (108-05-4), **65**, 135